AF371497

Lecture Notes in Electrical Engineering

Volume 101

Vadim Vasyukevich

Asynchronous Operators of Sequential Logic: Venjunction & Sequention

Digital Circuit Analysis and Design

 Springer

Dr. Vadim Vasyukevich
Lokomotives iela 56-93
LV-1057 Riga
Latvia
E-mail: vadim.v@balticom.lv

ISBN 978-3-642-21610-7   e-ISBN 978-3-642-21611-4

DOI 10.1007/978-3-642-21611-4

Lecture Notes in Electrical Engineering   ISSN 1876-1100

Library of Congress Control Number: 2011929655

Typeset & Cover Design: Scientific Publishing Services Pvt. Ltd., Chennai, India.

Printed on acid-free paper

9 8 7 6 5 4 3 2 1

springer.com

Preface

"What is to be done?"
Nikolay Chernyshevsky

First of all let me comply with A.L. Fradkov[1]: "Overall majority of science areas today look like a flowering meadow after a huge herd of animals had a walk on it. Grass is largely eaten and "juiciest peaces" are eroded up to the ground. Some areas are strolled upon numerous times. Here and there grass is only rumpled, but it would take an enormous effort to eat it, and the result would be so hardly noticeable that big animals are already seeking new pastures over many years..."

Now in essence.

This manuscript is dedicated to new mathematical instruments assigned for logical modeling of the memory of digital devices. The case in point is logic-dynamical operation named venjunction and venjunctive function as well as sequention[2] and sequentional function. Venjunction and sequention operate within the framework of sequential logic. In a form of corresponding equations, they organically fit analytical expressions of Boolean algebra. Thus, a sort of symbiosis is formed using elements of asynchronous sequential logic on the one hand, and combinational logic on the other hand. Thence, a common denomination is asynchronous logic.

A peak of publications on the asynchronous logic came in the middle of the 1990s. In spite of undertaken efforts and progress observed in this direction, it must be confessed that properly appreciable results are not achieved. At least not such, as have been expected. Together with this, relevance and claiming of asynchronous circuit as an object of science interest does not raise doubts. But it is evident that priority of investigations is displaced to programming and verification aspects as well as to the practical field of technology and technical engineering.

A wealth of methods, ways and applications developed for solving various problems related with digital circuits allow establishing the following.

- Asynchronous circuit analysis and design is a complicated and heterogeneous problem.
- There are no universal generally accepted approaches.

---

[1] Article "How to publish a good article and to reject a bad one. Notes of a reviewer" Available in Russian at: http://www.ipme.ru/ipme/labs/ccs/alf/f_at03r.pdf.

[2] Not to be confused with "sequent", used for the sequent calculus.

- All models and methods, as well as their applications, are not devoid of disadvantages.
- New ideas are desirable and required.

In the light of established situation, a curious thought arises: may be finite-automaton apparatus as being monopoly mathematical instrument for sequential circuits is the cause of intractable problems, because its scientific resource is exhausted.

Indeed?!

Foreword

In the beginning was Boolean algebra. Named in honor of George Boole, this algebra gave symbolic expressions for a binary logic. So the beginning relates to the 1840s.

Development of the theoretical grounds for Boolean algebra's digital applications has been basically completed in the 1950s. At that time mathematical instruments for representation, transformation and minimization of logical operations were ready

Boolean algebra constitutes a mathematical basis for digital circuits. It is known, how relatively easy Boolean equation is translated into logical circuit. Inverse transformation is performed without difficulties as well. In other words, a principle of one-to-one correspondence is on hand. Boolean expressions play the role of a natural analytical model of combinational digital circuits.

In contrast with combinational logic, there are no easy ways to represent behavior of logical circuits with feedbacks in an analytical form; tables and graphs are in use. Circuit's logic language is not in accord with state-transition expressions. Therefore, constructing of sequential circuit is a complicated, many-staged and heterogeneous procedure. As to inverse transformation, present analysis does not give formal methods for this problem. The corresponding solution is available only for simple circuits with memory elements.

The problem is how to formalize operation of memory devices mathematically in terms of binary logic, and how to get analytical model which enables to adequately reflect functionality of these devices. In this connection it is proposed to solve the mentioned problem by means of asynchronous operators, named venjunction and sequention. Under conditions that existing delays are unknown, algebraic expressions obtained on the basis of these operations one-to-one correspond to the modeled circuit. Thus, relation between asynchronous sequential circuit and its model becomes similar to relation between combinational circuit and expressions of Boolean algebra. As a result new analytical possibilities appear and some problems could be solved in easier way. Asynchronous operators may be found to be not only useful, but productive as well.

The book contains initial concepts, fundamental definitions, statements, principles and rules needed for theoretical justification of the new mathematical apparatus and its validity for asynchronous logic. Asynchronous operators named venjunctor and sequentor are designed for practical implementation. These basic elements are assigned for realizing of memory function in sequential circuits. It is important that in general case memory depth theoretically is not restricted.

The author develops original approach, whose superiority is in homogeneity of mathematical expressions. This approach does not substitute or copy the existing procedures; as well it does not solve all problems. Obtained results can be used independently or together with present methods.

Formed sequential logic essentially is classified as switching logic, which falls into category of algebraic logics. For better understanding reader should know Boolean algebra, set theory, and foundations of digital devices.

Present research work is the final stage of generalization and systematization of all those ideas and investigations, author's interest to which alternately flashed up and faded over many years and for various reasons until formed "critical mass", and all findings were arranged definitively as a mathematical basis of a theory appropriately associated under a common theme – asynchronous sequential logic.

Acknowledgments. I am grateful to everybody who did not bother me during writing of this book. Thanks also to my friends who from time to time distracted me with important and trivial dealings that helped me to preserve my potency and peace of mind. My special thanks to my beloved daughter Katerina for her participation in editing of manuscript, for her help with translation from Russian and invaluable moral support.

Contents

Chapter 1
Venjunction

Abstract. The first chapter is entirely dedicated to venjunction, which is represented as a logic-dynamical operation of asynchronous sequential logic. This operation is being thoroughly examined from all angles, such as: prerequisites of appearing, particularities of the implied time, necessary definitions, methods of analytical and graphical representations, enumeration of two-variable functions, and finishing by basic features of venjunctions in connection with operations of Boolean algebra. In the beginning of this section one will find clarifications of terminology to avoid incomprehension, and assure continuity between generally accepted notions and innovations. This to one or another degree refers to the following notions: format of a binary set, asynchronous sequence, logical switchings, moments and background of switchings, sequence of logical switchings, switching and venjunctive functions, venjunctive complete form, graph of venjunctive function, cyclic graph of switchings, and venjunctive representation of indeterminacy.

1.1 Binary Sets and Sequences

Let us assume that $[x_1\ x_2\ \ldots\ x_n]$ is an ordered collection consisting of all elements of a set $\{x_1, x_2, \ldots x_n\}$. These elements are binary variables with logical values 0 or 1. By replacing the variables with their values a binary combination is formed. Generally this combination can be considered as a word in the alphabet $\{0, 1\}$, and therefore termed binary set.

In the context of Boolean algebra, a set is said to be binary, if it contains logical zero (0) and unity (1) signs as values of certain Boolean variables. For example, variables $\{x_1, x_2, x_3, x_4, x_5\}$ with values $x_1=1$, $x_2=1$, $x_3=0$, $x_4=1$, $x_5=0$ form the binary set [11010]. Here, the first in order unit determines a value of the variable x_1 because it occupies the first position in the corresponding set. Similarly, second unit associates with the second symbol of the set, that is, x_2. And so on, up to zero value of the last, fifth variable x_5. Thus, there is positional correspondence between a binary set and a sequence of symbols located in a set of Boolean variables.

1.1.1 Format of Binary Set

For realization of the positional principle established above, the concept of a format of variables is entered. The format is represented in the form of a set of variables, the sequence of which predetermines values of these variables in the

V. Vasyukevich: Asynchronous Operators of Sequential Logic, LNEE 101, pp. 1–24.
springerlink.com   © Springer-Verlag Berlin Heidelberg 2011

appropriate binary set. Let's explain on an example. In view of the coordination of positions at values $x_1=1$, $x_2=1$, $x_3=0$, $x_4=1$, $x_5=0$ the appropriating binary set [11010] is presented in a format of a set of variables, or more simple, – in the format $[x_1\, x_2\, x_3\, x_4\, x_5]$. The record is allowed:

$$[x_1\, x_2\, x_3\, x_4\, x_5] = [11010]. \tag{1.1}$$

That same binary set, but formalized in a different format $[x_3\, x_2\, x_1\, x_5\, x_4]$, corresponds to values $x_3=1$, $x_2=1$, $x_1=0$, $x_5=1$, $x_4=0$. As a result of the change of formats other equality, different from the previous one, is obtained:

$$[x_1\, x_2\, x_3\, x_4\, x_5] = [01101]. \tag{1.2}$$

Thus, it is possible to consider that binary sets are positioned in accordance with a format of variables. Thanks to the format, binary sets are able to represent Boolean variables in an explicit form.

1.1.2 Binary Sequences

Unlike set a binary sequence represents values only for single variable, but not for whole set of them. So the sequence $\langle 11010 \rangle$ means that a certain variable x is presented by five values fixed during time, since the moment t_1 and finishing t_5. Point in time t_1 corresponds to the unity value $x(t_1) =1$, point t_2 – to $x(t_2) =1$, and further: $x(t_3) = 0$, $x(t_4) =1$, and $x(t_5) = 0$. Time depending variable $x(t)$ forms the sequence $\langle x(t_1),\ x(t_2),\ x(t_3),\ x(t_4),\ x(t_5) \rangle$, which determines a format for the proper construction of binary sequences, answering $\langle x \rangle$. In considered case a following equality takes place:

$$\langle x(t_1)\, x(t_2)\, x(t_3)\, x(t_4)\, x(t_5) \rangle = \langle 11011 \rangle. \tag{1.3}$$

Thus, a binary sequence is positioned in time. Each value of a corresponding variable is caused by a moment of time when this variable appears in an explicit form: 0 or 1.

1.1.3 Asynchronous Sequences

A distinctive feature of asynchronous binary sequence is that it does not imply any external control of time points t_1, t_2, t_3, and so on. There is no external synchronizer which would set these points, placing the certain marks on the axis of time during which variable $x(t)$ is traced. In other words, the binary sequence of a format $\langle x(t_1)\, x(t_2)\, x(t_3)\ \ldots\ x(t_\mathrm{m}) \rangle$ is called asynchronous if a temporal sequence $\langle t_1\, t_2\, t_3\ \ldots\ t_\mathrm{m} \rangle$ is asynchronous.

1.1.4 Intersection of Sets and Sequences

Followings after each other binary sets, as well as asynchronous sequences collected together, in fact, represent how and which variables change their value over a fixed time. In addition, these representations are orthogonal; they overlap. Therefore binary sets and sequences are able to be jointly displayed within the framework of a common table (Table 1.1).

Table 1.1 Binary sets and asynchronous sequences

		x_1	x_2	x_3	x_4
t_1	X_1	0	0	0	1
t_2	X_2	0	1	0	1
t_3	X_3	1	1	0	1
t_4	X_4	1	1	0	0
t_5	X_5	1	1	1	0

This table is structured as follows. There are horizontally located binary sets: $X_1=$ [0001], $X_2=$ [0101], $X_3=$ [1101], $X_4=$ [1100], $X_5=$ [1110]. They are fixed in moments of time t_1, t_2, t_3, t_4, t_5 respectively. All sets are presented in the format $[x_1\ x_2\ x_3\ x_4]$ composed of four binary variables. To these variables answer vertically located sequences: $x_1(t) = \langle 00111 \rangle$, $x_2(t) = \langle 01111 \rangle$, $x_3(t) = \langle 00001 \rangle$, $x_4(t) = \langle 11100 \rangle$, which are presented in the format $\langle X_1\ X_2\ X_3\ X_4\ X_5 \rangle$ composed of binary sets. Asynchronous property of binary sequences is caused by asynchronous temporal sequence $\langle t_1, t_2, t_3, t_4, t_5 \rangle$.

1.2 Logical Switchings

Attention is drawn to that data presented in Table 1.1 can be examined as a series of logical switchings. Thus, under switching is meant changing of value of one or another binary variable from a logical zero to logical unity or from unity to zero: 0/1 or 1/0 respectively. Switchings are carried out in the following order. First it concerns a variable x_2. It changes the value from logical zero to unity, because at the moment of time t_1 equality $x_2(t_1) = 0$ is observed, and at the next moment $t_2 - x_2(t_2) = 1$. A switching on hand is $x_2 = 0/1$, it causes change of a binary set $X_1=$ [0001] to $X_2=$ [0101]. Further switchings $x_1 = 0/1$, $x_4 = 1/0$ and $x_3 = 0/1$, which meet the moments of time t_3, t_4 and t_5 with binary sets $X_3=$ [1101], $X_4=$ [1100] and $X_5=$ [1110] respectively, take place.

1.2.1 Moments of Switchings

Moments of switchings are tied to the moments of time when the values of variables are fixed. This obvious circumstance admits different interpretations because of binary variable behavior directly at the time of logical switching. Therefore, eliminating possible collisions and without losing a generality, we believe that any asynchronous sequence $x(t)$ in the neighborhoods of the moment of time t_j obeys the following rules:

$$x(t < t_j) = x(t_{j-1}), \tag{1.4.1}$$

$$x(t \geq t_j) = x(t_j) . \tag{1.4.2}$$

Until the time t_j variable x does not change its value. At the moment t_j switching occurs and variable takes on another value that meets $x(t_j)$. This new value holds constant until the moment of the next switching. Concretely, if $x(t_{j-1}) = 0$, but $x(t_j) = 1$, then at the moment of time t_j variable x is switched from logical zero to unity: $x = 0/1$. If $x(t_{j-1}) = 1$ but $x(t_j) = 0$, switching $x = 1/0$ takes place.

Within the framework of the offered concept an imaginary, in sense of "pseudo", switchings of type $x = 0/0$ and $x = 1/1$ are allowed as well. As this takes place, equality $x(t_{j-1}) = x(t_j)$ according to which a variable does not change its value at the moment t_j is obeyed. Real switching is always accompanied by an inequality $x(t_{j-1}) \neq x(t_j)$.

1.2.2   Background for Switchings

If logical switching is an action, the role of background is rightly assigned to pseudoswitchings. An example may clarify this. When passing from binary set X_2 to set X_3 (Table 1.1) the following happens. Variable x_1 changes its value from logical zero to unity. At the same time other variables do not change their own values and thus create a peculiar background for switching $x_1 = 0/1$. Background switches at the moment of time t_3 are formally displayed in the form of pseudoswitchings: $x_2 = 1/1$, $x_3 = 0/0$ and $x_4 = 1/1$.

A background in itself is meaningful only in a combination with switching, for example "*switching $x_1 = 0/1$ on the background $x_2 = 1/1$*". A simplified record is allowed also: "*switching $x_1 = 0/1$ on the background $x_2 = 1$*".

1.2.3   Rules for Switchings

Transition from any binary set to the following occurs due to switchings of variables. Order of these switchings is caused by the certain rules.

1. All switchings are asynchronous. They are separated in time in such a manner that during each moment a switching of only one variable can happen.
2. Each switching generates a new binary set. This set differs from previous by value of a single one variable whose switching, in fact, fixes the moment of time when binary set changes.

1.3 Methods for Representation of Variables Collections

Judging from configuration of Table 1.1 as an example, its content can be expressed in analytical form. For this purpose it is enough to take advantage of the following methods.

1.3.1 Representation by a Sequence of Binary Sets

For example, the sequence looks as follows: $\langle[0001]\ [0101]\ [1101]\ [1100]\ [1110]\rangle$. Here, all values of binary variables are in agreement with the following denotations:

$$[0001] = [\langle x_1(t_1)\ x_2(t_1)\ x_3(t_1)\ x_4(t_1)\rangle] = X_1, \tag{1.5.1}$$

$$[0101] = [\langle x_1(t_2)\ x_2(t_2)\ x_3(t_2)\ x_4(t_2)\rangle] = X_2, \tag{1.5.2}$$

$$[1101] = [\langle x_1(t_3)\ x_2(t_3)\ x_3(t_3)\ x_4(t_3)\rangle] = X_3, \tag{1.5.3}$$

$$[1100] = [\langle x_1(t_4)\ x_2(t_4)\ x_3(t_4)\ x_4(t_4)\rangle] = X_4, \tag{1.5.4}$$

$$[1110] = [\langle x_1(t_5)\ x_2(t_5)\ x_3(t_5)\ x_4(t_5)\rangle] = X_5. \tag{1.5.5}$$

For a general case asynchronous sequence of binary sets is given by the following formal expression:

$$\langle[X]\rangle = \langle X_1\ X_2\ X_3\ \dots\ X_{m-1}\ X_m\rangle. \tag{1.6}$$

Here "m" is a length of sequence, which is caused by number of points used for fixing the values of Boolean variables.

1.3.2 Representation by a Set of Asynchronous Sequences

For example, the set looks as follows: $[\langle00111\rangle\ \langle01111\rangle\ \langle00001\rangle\ \langle11100\rangle]$. Here, values assigned to binary variables correspond to the columns of table:

$$\langle00111\rangle = x_1(t) = \langle x_1\rangle, \tag{1.7.1}$$

$$\langle01111\rangle = x_2(t) = \langle x_2\rangle, \tag{1.7.2}$$

$$\langle00001\rangle = x_3(t) = \langle x_3\rangle, \tag{1.7.3}$$

$$\langle11100\rangle = x_4(t) = \langle x_4\rangle. \tag{1.7.4}$$

In the general case a set of binary sequences is presented by the expression:

$$[\langle X\rangle] = [\langle x_1\rangle,\ \langle x_2\rangle,\ \langle x_3\rangle\ \dots\ \langle x_{n-1}\rangle\ \langle x_n\rangle], \tag{1.8}$$

where "n" is the cardinality of a set of involved Boolean variables.

1.3.3 Representation by a Sequence of Logical Switchings

Switchings of variables occur in accordance with priority, which is defined in the following sequence:

– $\langle (x_2 = 0/1)\,(x_1 = 0/1)\,(x_4 = 1/0)\,(x_3 = 0/1)\rangle$.

With due account of the variables formatting logical switchings naturally present in equality:

$$\langle x_2\,x_1\,x_4\,x_3 \rangle = \langle (0/1)\,(0/1)\,(1/0)\,(0/1)\rangle, \tag{1.9}$$

whose components satisfy the following relations:

$$x_2 = 0/1 = x_2(t_1)\,/x_2(t_2), \tag{1.10.1}$$

$$x_1 = 0/1 = x_1(t_2)\,/x_1(t_3), \tag{1.10.2}$$

$$x_4 = 1/0 = x_4(t_3)\,/x_4(t_4), \tag{1.10.3}$$

$$x_3 = 0/1 = x_3(t_4)\,/x_3(t_5). \tag{1.10.4}$$

1.4 Switching Function – Venjunction

On the basis of logical switchings a switching function is built. For its realization an operation called venjunction is involved (*whenjunction* in [1]), [2].

1.4.1 Operation of Venjunction

Verbal expression written as "*switching … on the background …*" has a formalized mathematical representation. For this purpose a special operation named venjunction and designated by sign " $\angle$ " is intended. This sign is asynchronous operator, which links binary variables. As an example, for pair of variables x and y the expression "*switching x on the background y*" is presented by formula $x \angle y$.

Venjunction is an asymmetrical logic dynamical operation. It takes into account the values of variables not only current, but also their previous moment of time. The operation is asymmetrical in the sense of inequality:

$$x \angle y \neq y \angle x . \tag{1.11}$$

Switching x on the background y and reverse switching y on the background x are different actions, not compatible in time of their realization. Venjunctions associated with these switchings are "mirrors" in the relation to one another.

1.4.2 Switching Function

As a switching is named a binary function if it essentially depends on switchings of arguments – Boolean variables. The corresponding changes are considered owing to the venjunction, logical truth of which is determined by the value of binary variable z in equation:

$$z = x \angle y . \tag{1.12}$$

This formula serves as a basis for construction of switching function which components are tabulated (Table 1.2) in view of the format of Table 1.1.

Table 1.2 Components of switching function.

		x	y	z
t_{j-1}	X_{j-1}	$x(t_{j-1})$	$y(t_{j-1})$	$z(t_{j-1})$
t_j	X_j	$x(t_j)$	$y(t_j)$	$z(t_j)$

Variables x and y, as being arguments, are fixed at the moments of time t_{j-1} and t_j. Values of these variables $x(t_{j-1})$, $y(t_{j-1})$ and $x(t_j)$, $y(t_j)$ belong to binary sets X_{j-1} and X_j respectively.

Switching function is presented by a variable z determined at the moment of time t_j with help of the function of the following type:

$$z(t_j) = F(x(t_j), y(t_j), x(t_{j-1}), y(t_{j-1}), z(t_{j-1})). \tag{1.13}$$

Taking into account switchings of variables x and y a dependence obtained is adequately displayed by the following formula:

$$z(t_j) = F(x(t_{j-1})/x(t_j), y(t_{j-1})/y(t_j), z(t_{j-1})). \tag{1.14}$$

1.4.3 Venjunction as Function

Operation of venjunction implies that the truth of the respective function, which is expressed by the logical unity $z(t_j) = 1$, is caused by switching $x(t_{j-1})/x(t_j) = 0/1$ on the background $y(t_{j-1})/y(t_j) = 1/1$. Corresponding values of components of the switching function are presented in Table 1.3.

Table 1.3 Truth table for venjunction.

		x	y	z
t_{j-1}	X_{j-1}	0	1	J
t_j	X_j	1	1	1

When a logical switching $x = 0/1$ on the background $y = 1/1$ takes place, the operation of venjunction $x \angle y$ ensures the switching $z = J/1$, where $J \in \{0,1\}$. Symbol J characterizes indeterminacy, which in this case results from two abilities: $z = 0/1$ or $z = 1/1$. Equality $z(t_j) = 1$ indicates that switching function attains a logical true value. A false, that is zero value $z(t_j) = 0$, accompanies switchings which answer data of Table 1.3. These cases are $x = 0/1$, $x = \overline{0/1}$ and $y = \overline{1/1}$.

From equality $z(t_{j-1}) = J$ it follows that a value assigned to variable $z(t_{j-1})$ does not influence on $z(t_j)$. Therefore, for venjunction as for switching function it is characteristic the following dependence:

$$z = z(t_j) = F(x(t_{j-1}), y(t_{j-1}), x(t_j), y(t_j)). \tag{1.15}$$

In the cause of switching of variables x, y and z, venjunctive function is able to generate various binary combinations. All of them are collected in ten tables of Table 1.4. Each table is constructed in view of the format accepted for Table 1.3.

Table 1.4 Switchings caused by venjunction (Table 1.3 format).

0	1	0	0	0	0	1	1	0	1	1	1	1	0	0
1	1	1	1	0	0	0	1	0	0	1	0	0	0	0
1	0	0	0	0	0	1	1	0	1	1	1	0	1	0
1	1	0	0	1	0	1	0	0	1	0	0	0	0	0

Given analytical calculations serve as a basis for a number of assertions, by means of which functional possibilities of venjunctive dependence are characterized in full measure. In terms of logical switchings these assertions are formulated as follows:

- if $x = 0/1$ on the background $y = 1/1$, then $x \angle y = 0/1$;
- if $y = 0/1$ on the background $x = 1/1$, then $x \angle y = 0/0$;
- if $x = 1/0$ on the background $y = 1/1$, and $z(t_{j-1}) = 1$, then $x \angle y = 1/0$;
- if $x = 1/0$ on the background $y = 1/1$, and $z(t_{j-1}) = 0$, then $x \angle y = 0/0$;
- if $y = 1/0$ on the background $x = 1/1$, and $z(t_{j-1}) = 1$, then $x \angle y = 1/0$;
- if $y = 1/0$ on the background $x = 1/1$, and $z(t_{j-1}) = 0$, then $x \angle y = 0/0$;
- if a background is $x = 0/0$ or $y = 0/0$, then $x \angle y = 0/0$.

Formally, considering pseudoswitchings the following expressions should be added:
- if $x = 1/1$ and $y = 1/1$, but $z(t_{j-1}) = 1$, then $x \angle y = 1/1$;
- if $x = 1/1$ and $y = 1/1$, *but* $z(t_{j-1}) = 0$, then $x \angle y = 0/0$;
- if $x = 0/0$ or $y = 0/0$, then $x \angle y = 0/0$.

1.5 Venjunction in Comparison with Conjunction

Judging from Table 1.3 function $z = x \angle y$, as well as conjunction $x \wedge y$, produces unity value, if and as a result of unity values of input variables: $x = y = 1$. But regarding $x \angle y = 1$ there is one more sufficient condition, namely: logical unity setting for x must occur at the moment when unity value $y = 1$ is already fixed. That is, logical switching $x = 0/1$ must be realized on the background of a steady value $y = 1$. An essence of operation is explained by a diagram in Fig. 1.1.

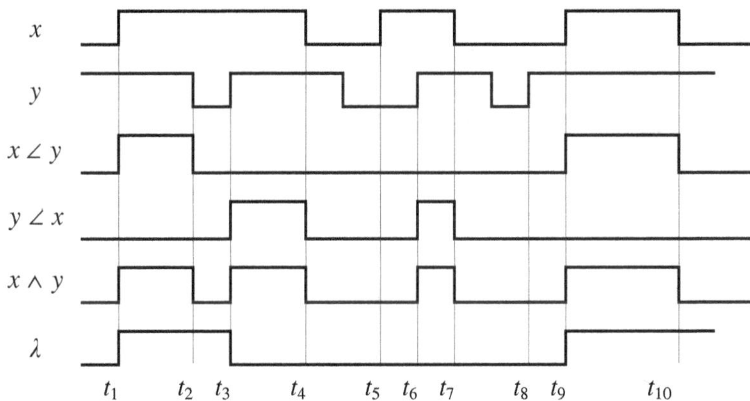

Fig. 1.1 Venjunction and conjunction; comparative pulse diagram

According to the diagram, at the moments of time t_1 and t_9, when $y = 1$, and variable x changes the value in connection with its switching from logical zero to unity (0/1), a resulting function reduces to unity. On a contrary due to $y = 0$, switching $x = 0/1$ at the moment t_5 does not produce an effect on the graph of venjunction $x \angle y$. Unity of the corresponding function is maintained by the unity values of arguments. When a variable x or y reduces to zero, function value becomes zero. In the diagram it is shown by the moments of time t_2 and t_{10}, which are connected with switchings $y = 1/0$ and $x = 1/0$ respectively.

A mirror function, as being venjunction $y \angle x$, behaves similarly. It takes on a unity value at the moments t_3 and t_6 when switching $y = 0/1$ happens on the background $x = 1$. Function becomes zero at the moments of time t_4 and t_7. According to diagram, mentioned above "mirroring" is expressed in a form of the inequality $(y \angle x) \neq (x \angle y)$ with the following relation: if $x \angle y = 1$, then $y \angle x = 0$ and vice versa. By such a manner a venjunction function demonstrates its sensitivity to the variables interchanging.

Behaviors of venjunctions $x \angle y$ and $y \angle x$ in Fig. 1.1 are displayed in comparison with conjunction $x \wedge y$. So it is seen, that each venjunction can be interpreted as a "truncated" conjunction. Actually, while it is realized unity switching

$x = 0/1$ on the background $y = 1$, a coincidence between values of functions $x \angle y$ and $x \wedge y$ is observed. In the opposite case, when signal y produces a unity value on the background $x = 1$, a conjunction $x \wedge y$ as to its value does not differ from the mirror venjunction $y \angle x$.

Thus, function of conjunction unites connected with it mirror venjunctions by means of the formula:

$$x \wedge y = (x \angle y) \vee (y \angle x) . \tag{1.16}$$

The given expression formalizes the rule of disjunctive expansion of a conjunction into two venjunctions as components. Or otherwise (reverse "reading"), it is a rule of merging of venjunctions.

1.6 Methods of Venjunction Definition

Venjunction, or venjunctive function, is defined in accordance with dependence $z = x \angle y$. Without leaving the framework of Boolean algebra, we believe that operation $x \angle y$ generates a function with a range of values $\{0, 1\}$. Below there are presented various kinds of methods, which are applicable for a venjunction defining.

1.6.1 Definition Based on Binary Sets and Sequences

- *if* $[x\, y] = [11]$ with condition $\langle [x\, y] \rangle = \langle [01]\, [11] \rangle$, then $x \angle y = 1$;
- *if* $[x\, y] = [11]$ with condition $\langle [x\, y] \rangle = \langle [10]\, [11] \rangle$, then $x \angle y = 0$;
- if $x = 0$ or $y = 0$ ($x \vee y = 0$), then $x \angle y = 0$.

1.6.2 Definition with Involving an Indeterminate Value

Uncertain value of a variable takes place in the case, when this variable is not uniquely defined. To this variable can be assigned unity as well as zero value. Uncertainty in the form of a symbol J is convenient for using, if value of a variable is not known, or the choice of value is free or indifferent (depending on a context). Formally, J is a third, alongside with 1 and 0, value of a binary variable. However, this value is not self-sufficient because of $J \in \{1, 0\}$.

Based on the foregoing the function of venjunction is defined as follows:

- if $\langle [x\, y] \rangle = \langle [01]\, [11] \rangle$, then $x \angle y = 0/1$;
- if $\langle [x\, y] \rangle = \langle [10]\, [11] \rangle$, then $x \angle y = 0/0$;
- if $\langle [x\, y] \rangle = \langle [JJ]\, [J0] \rangle$, then $x \angle y = J/0$;
- if $\langle [x\, y] \rangle = \langle [JJ]\, [0J] \rangle$, then $x \angle y = J/0$.

1.6.3 Definition with Involving a Conjunction

- if $x \wedge y = 1$ after a moment $x = 0/1$, then $x \angle y = 1$;
- if $x \wedge y = 0$ or $y \angle x = 1$, then $x \angle y = 0$.

1.6.4 Definition with Involving Ancillary Binary Variable

On the above diagram (Fig. 1.1) alongside with variables which serve as arguments for venjunction function, there is the auxiliary binary variable designated by a symbol λ. Values of this variable are defined as follows:

- $\lambda = 1$ after a moment $x = 0/1$ when $y = 1$;
- $\lambda = 0$ after a moment $y = 0/1$ when $x = 1$.

Moments of switching $\lambda = 0/1$ are points in time t_1 and t_9; switching $\lambda = 1/0$ occurs at the moment t_3. By using the auxiliary variable venjunction is defined through a conjunction in accordance with the following formula:

$$x \angle y = \lambda \wedge x \wedge y .$$
$$(1.17)$$

Similar expression for mirror venjunction differs only by the value of the auxiliary variable:

$$x \angle y = \overline{\lambda} \wedge x \wedge y .$$
$$(1.18)$$

1.6.5 Not Formal Definition: Verbal "Reading" of Venjunction

There are a variety of options. For example, the detailed (developed) record:

- the function $z = x \angle y$ takes on a unity value in the case of switching of variable x from zero to unity at the constant unity value of variable y. Established unity value of function remains until the moment when x or y reduces to zero.

Truncated variant:
- the function $z = x \angle y$ becomes unity at switching $x = 0/1$ on the background $y = 1$; it stays in unity until variables x and y retain their unity values.

Example of minimized reading of venjunction:
- $z = 1$ from the moment $x = 0/1$ on the background $y = 1$, and up to $x = 0$ or $y = 0$.

As required the opportunity exists, and it is possible to introduce venjunction into algorithmic constructions with formulation: *"if x when y, then…"*

1.7 Truth Tables for Venjunction

Truth of the function $z = x \angle y$ is caused by values and switchings of variables x, y, and it can be established with the help of the corresponding tables. Table 1.5 represents the following dependence:

$$z(t_{j-1})/z(t_j) = F(x(t_{j-1})/x(t_j), y(t_{j-1})/y(t_j)), \qquad (1.19)$$

according to which a function (z) and its arguments (x, y) take the form of logical switchings.

Left column of the table contains switchings of the binary variable x. Switchings of the binary variable y are placed in upper row. In the squares located at the intersection of columns and rows, switchings of resulting variable z are presented. For example, if $x = 0/1$ and $y = 1/1$, then $z = 0/1$. Indetermination of $z = J/J$ caused by pseudoswitchings $x = 1/1$ and $y = 1/1$ means that in the present context two results $z = 0/0$ and $z = 1/1$ can be equally supposed.

If switching values contradict the rules accepted for the binary variables, the corresponding squares for z remain not filled. For example, the combination of switchings $x = 0/1$ and $y = 1/0$ is not subject to examination. This combination is considered inadmissible because at once two variables simultaneously change their values that is forbidden a priori (see Sect. 1.2.3).

Table 1.5 Venjunctive dependence (Eq. 1.19).

		y			
		1/1	0/1	1/0	0/0
	1/1	J/J	0/0	J/0	0/0
x	0/1	0/1	-	-	0/0
	1/0	J/0	-	-	0/0
	0/0	0/0	0/0	0/0	0/0

Table 1.5 contains all switchings concerned with the operation of venjunction, including those which are typical for a venjunctive function. In view of accepted above definitions (see Sect. 1.6) values of this function are presented by the second component of the switching $z(t_{j-1})/z(t_j)$, that means $z(t_j)$. The corresponding functional dependence is represented by formula:

$$z(t_j) = F(x(t_{j-1})/x(t_j), y(t_{j-1})/y(t_j)). \qquad (1.20)$$

On the basis of given dependence a truth table of venjunction is constructed (Table 1.6).

Table 1.6 Venjunctive dependence (Eq. 1.20).

		y			
		1	0/1	1/0	0
	1	-	0	0	-
x	0/1	1	-	-	0
	1/0	0	-	-	0
	0	-	0	0	-

Another way of venjunction representation is caused by functional dependence of the following type:

$$z(t_2) = F(\langle [x(t_1)\, y(t_1)]\, [x(t_2)\, y(t_2)] \rangle). \tag{1.21}$$

Accordingly, truth values of function are determined in Table 1.7. This table differs from Table 1.5 and Table 1.6 in that its function is defined on sequences of binary sets and not on the switchings of binary variables. For example, instead of the switching $x = 0/1$ and constant value $y = 1$ two binary sets $[x\,y] = [01]$ and $[x\,y] = [11]$ in the sequence $\langle [01]\,[11] \rangle$ are used.

Table 1.7. Venjunctive dependence (Eq. 1.21).

t_1		t_2		
X_1		X_2		
x	y	x	y	z
0	1	1	1	1
1	1	1	0	0
1	0	0	0	0
0	0	1	0	0
1	0	1	1	0
1	1	0	1	0
0	1	0	0	0
0	0	0	1	0

At the moment of time t_1 a binary set $X_1 = [x(t_1)\, y(t_1)]$ acts. At the next moment t_2 a set $X_2 = [x(t_2)\, y(t_2)]$ interchanges the previous one. A function is assigned by the values of z at the time t_2, that is $z(t_2)$. Logical switchings of variables x and y are presented in the form of sequences of two sets: $\langle X_1\, X_2 \rangle$. These sets, as well as their corresponding values of $z = z(t_2)$, are located line by line (row-wise).

Table 1.7 is constructed in such a manner that step-by-step shifting from upper row of binary sets to the lower row forms the following sequence:

- $\langle$[01] [11]$\rangle$, $\langle$[11] [10]$\rangle$, $\langle$[10] [00]$\rangle$, $\langle$[00] [10]$\rangle$, $\langle$[10] [11]$\rangle$, $\langle$[11] [01]$\rangle$, $\langle$[01] [00]$\rangle$, $\langle$[00] [01]$\rangle$.

The sequence can be adequately minimized by removing all repeated sets. Then:

- $\langle$[01] [11] [10] [00] [10] [11] [01] [00] [01]$\rangle$.

In this case all switchings which are typical for two variables are presented. This feature serves as a basis for constructing yet another table (Table 1.8). This master table aside from a basic venjunction $x \angle y$ incorporates all its modifications which are generated by permutation and logical inversion of the variables x and y. Letter V denotes venjunction so that:

$$V_1 = x \angle y, \quad V_2 = \overline{y} \angle x, \quad V_3 = \overline{x} \angle \overline{y}, \quad V_4 = x \angle \overline{y},$$
$$V_5 = y \angle x, \quad V_6 = \overline{x} \angle y, \quad V_7 = \overline{y} \angle \overline{x}, \quad V_8 = y \angle \overline{x}.$$

Table 1.8 Master truth table for venjunctions.

		x	y	V_1	V_2	V_3	V_4	V_5	V_6	V_7	V_8
t_1	X_1	0	1	0	0	0	0	0	J	0	J
t_2	X_2	1	1	1	0	0	0	0	0	0	0
t_3	X_3	1	0	0	1	0	0	0	0	0	0
t_4	X_4	0	0	0	0	1	0	0	0	0	0
t_5	X_5	1	0	0	0	0	1	0	0	0	0
t_6	X_6	1	1	0	0	0	0	1	0	0	0
t_7	X_7	0	1	0	0	0	0	0	1	0	0
t_8	X_8	0	0	0	0	0	0	0	0	1	0
t_9	X_9	0	1	0	0	0	0	0	0	0	1

In the master table venjunctive dependences are presented in the function form: $z(t) = F(\langle[X]\rangle)$.

The order, in which binary sets $X = [x\ y]$ are arranged, is caused by the format:

$$\langle[X]\rangle = \langle[X_1 X_2 X_3 X_4 X_5 X_6 X_7 X_8 X_1]\rangle. \tag{1.22}$$

The values of each venjunctive function are displayed in the form of asynchronous sequence:

$$z(t) = \langle z(t_1)\, z(t_2)\, z(t_3)\, z(t_4)\, z(t_5)\, z(t_6)\, z(t_7)\, z(t_8)\, z(t_9) \rangle. \tag{1.23}$$

For example, truth value of the venjunction $V_2 = \overline{y} \angle x$ is characterized by a sequence: $z(t) = \langle 001000000 \rangle$.

1.8 Venjunctive Functions and Their Enumeration

Venjunctive functions in essence are switching functions; their analytical representation is inextricably linked with application of the operation named venjunction. Venjunctive function in the minimal form is an elementary venjunction of two variables. All logic functions of two variables are subject to the announced enumeration.

1.8.1 Venjunctive Complete Form

Venjunctive complete form is the universal formal way available for representing the logic functions having two variables. Such opportunity is a result of the informative part of Table 1.8, where each eight venjunctions becomes unity exclusively at one set of the binary sequence. It is notable that under this set all other venjunctions are reduced to zero. Therefore, it is not difficult to construct any of possible 2^8 functions by means of the Boolean disjunctive operation.

1.8.1.1 Venjunctive Form Completed by Disjunction

Common expression for the venjunctive complete form is given by the following:

$$\bigvee_{i} (k_i \wedge V_i), \; i \in \{1, 2 \dots 8\}. \tag{1.24}$$

Coefficient k regulates the sampling of such venjunctions that are chosen for representation of one or another function in its complete form. Representation takes place at unity value of k $(k - 1)$. In the opposite case $(k = 1)$ $(k = 0)$ the corresponding venjunction does not take part in formation of function.

For example, if venjunctions V_1, V_2, V_4, V_5, and V_6 are chosen, they form the function which in its complete form looks as follows:

$$z = (x \angle y) \vee (\overline{y} \angle x) \vee (x \angle \overline{y}) \vee (y \angle x) \vee (\overline{x} \angle y). \tag{1.25}$$

As it is seen venjunctive expression represented in complete form looks like disjunction of venjunctions.

In view of the rule established above (see Sect. 1.5, Eq. 1.16) for mirror venjunctions, the given formula is minimized to the compact expression:

$$z = x \vee (\overline{x} \angle y). \tag{1.26}$$

It is necessary to note that a compact form cannot include more than four venjunctions. Whole set of venjunctions contains four mirror pairs. Therefore, any fifth venjunction is a mirror in relation to existing venjunction, and together they generate a conjunction.

1.8.1.2 Venjunctive Form Completed by Conjunction

In an effort to enumerate the functions, another expression is also applicable. It is an alternative complete form, where logical operations of conjunction and negation are used in such a manner:

$$\bigwedge_i (\overline{k_i} \vee \overline{V_i}), \quad i \in \{1, 2 \ldots 8\}, \tag{1.27}$$

In this case chosen above six venjunctions (see Sect. 1.8.1.1) form a function represented by the following expressions:

$$z = \overline{V_1} \wedge \overline{V_2} \wedge \overline{V_4} \wedge \overline{V_5} \wedge \overline{V_6} = \overline{(x \angle y)} \wedge \overline{(\overline{y} \angle x)} \wedge \overline{(x \angle \overline{y})} \wedge \overline{(y \angle x)} \wedge \overline{(\overline{x} \angle y)}, \tag{1.28.1}$$

$$z = V_3 \vee V_7 \vee V_8 = (\overline{x} \angle \overline{y}) \vee (\overline{y} \angle \overline{x}) \vee (y \angle \overline{x}) = \overline{x} \wedge \overline{y} \vee y \angle \overline{x}. \tag{1.28.2}$$

Note
Venjunctive complete forms play the same role as full disjunctive normal form and full conjunctive normal form in Boolean algebra.

It is obvious that venjunctive complete form based on disjunctive operation is more suitable for enumeration in contrast with another form based on conjunctions. Conjunctive form is burdened with additional calculations, which are caused by logical negation.

1.8.2 Enumeration of Functions of Two Variables

General aim of the proposed enumeration is to obtain a full collection of binary functions with two variables. In the present context only venjunctive functions are implied, that is functions containing at least one operation of venjunction. The full list of the corresponding expressions is shown in Appendix A.

As a matter of convenience, functions are presented in compact form and, what is more, they are classified depending on the number of combined venjunctions, conjunctions and variables.

Operations that ensure obtaining of venjunctive compact forms are supported by transformations of mirror venjunctions and by the laws of Boolean algebra.

So by means of Boolean's disjunction every pair of mirror venjunctions is reduced to conjunction (see Sect. 1.5, Eq. 1.16). This transformation is reflected in the columns of Table 1.9.

In turn, using disjunctive rule, a pair of conjunctions is reduced to a binary variable in accordance with columns in Table 1.10.

Table 1.9 Transformation "venjunctions – conjunction

Venjunction mirror pairs	$x \angle y$	$\overline{x} \angle \overline{y}$	$x \angle \overline{y}$	$\overline{x} \angle y$
	$y \angle x$	$\overline{y} \angle \overline{x}$	$\overline{y} \angle x$	$y \angle \overline{x}$
Conjunctions	$x \wedge y$	$\overline{x} \wedge \overline{y}$	$x \wedge \overline{y}$	$\overline{x} \wedge y$

Table 1.10 Transformation "conjunctions – variable".

Pairs of conjunctions	$x \wedge y$	$\overline{x} \wedge \overline{y}$	$x \wedge \overline{y}$	$\overline{x} \wedge y$
	$x \wedge \overline{y}$	$\overline{x} \wedge y$	$\overline{x} \wedge \overline{y}$	$x \wedge y$
Variables	x	$\overline{x}$	$\overline{y}$	y

After minimizations performed on the basis of Tables 1.9 and 1.10, venjunctive functions appear in the form of compact expressions, where beside of venjunctions Boolean components such as conjunctions and binary variables are allowed.

Table 1.11 Combinability of venjunctions (v) with conjunctions (c) and variables (v).

	0	1 c	2 c	1 v	2 v
1 v	8	24	8	16	8
2 v	24	48	8	16	0
3 v	32	32	0	0	0
4 v	16	0	0	0	0

Table 1.11 contains quantitative data of all enumerated venjunctive functions: possible combinations and numbers of the forming components. This table establishes the following rules.

1. Maximum number of venjunctions is not more than four.
2. Maximum number of conjunctions and variables is not more than two.
3. Four venjunctions can not be supplemented by another component.
4. Only a single venjunction can be combined with two variables.
5. Three venjunctions can be combined only with one component, namely conjunction.

As a result of enumeration it is discovered that 243 functions (see App. A) satisfy the given conditions. Herewith 240 of them (see Table 1.11) depend on both variables. It is a great supplement to 16 functions which are available within the framework of the Boolean algebra. Three more functions, named truncated, are characteristic only for sequential logic.

1.9 Graphics of Venjunctive Functions

1.9.1 Graphical Representation of Switchings

Alongside with tabular form (see Sect. 1.7), a visual presentation of venjunctive functions can be made by using an adequate graphical constructions. In particular, data of Table 1.6 are in agreement with graphs of venjunction $z = (x \angle y)$ displayed in Fig. 1.2.

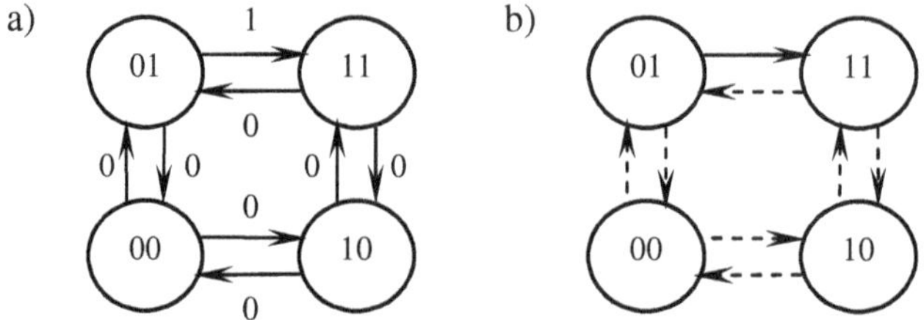

Fig. 1.2 Graphs of operation of venjunction

Nodes of the graphs are binary sets [01], [11], [10] and [00], presented in the format [x y]. Nodes are connected by arcs, which symbolize transitions between nodes. Arcs represent output transition from one value associated with arc beginning to another value associated with arrow.

Each transition is caused by the corresponding switching of variables. For example, a switching $x = 0/1$ on the background $y = 1$ is displayed by the arc, which comes from node [01] and reaches a node [11].

In Fig. 1.2a each edge-switching is accompanied by a digit, that is 1 (unity), because in response to the switching $x = 0/1$ the given function produces a value $z = 1$. Other switchings, as it is characteristic for venjunction, cause a zero value $z = 0$.

In Fig. 1.2b graph of venjunction is represented in the different form. Arcs are not burdened with function values 0 and 1. Instead of this, it is meant that every unbroken line of arc indicates $z = 1$ setting, and dotted line is associated with the value $z = 0$.

1.9.2 Graph of Cycles of Switchings

Aside from the previous graph (Fig 1.2) cyclic one is build on the basis of truth table. It is a sort of graphic copy of a truth table for venjunction (see Sect. 1.7, Table 1.5) looking like a cube with six cycles.

As example, the following venjunctive functions:

$$z = (x \angle y) \vee x \wedge \overline{y} , \tag{1.29.1}$$

$$z = (x \angle y) \vee (\overline{x} \angle y) \vee (x \angle \overline{y}) \vee \overline{x} \wedge \overline{y} . \tag{1.29.2}$$

are displayed in Fig. 1.3.

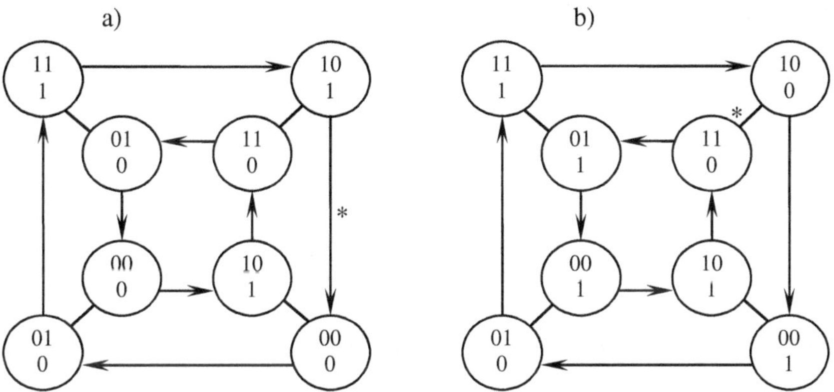

Fig. 1.3 Cyclic graphs of venjunctive functions: a) Eq. 1.29.1, b) Eq. 1.29.2

A cyclic graph represents venjunctive functions on sequences of input switchings. Here, just as input sets, output values are also placed in graph nodes.

For example, according to arc marked by asterisk "*" in Fig. 1.3a output value changes from 1 to 0 in the course of the switching $x = 1/0$ on the background $y = 0$.

Graph edges allow transitions in both directions. For example, edge marked in Fig. 1.3b characterizes switchings $y = 0/1$ and $y = 1/0$ on the background $x = 1$. These switchings do not change zero output $z = 0$.

By means of graphic image the connection between input switchings and output values that follows these switchings is traced most clearly. For example (Fig. 1.3a), input sequence [01], [11], [10], [11], [01] corresponds to output values 0, 1, 1, 0, 0 respectively.

1.9.2.1 Relationship with Venjunctive Complete Form

Switching graph of venjunctive function in essence is graphic images of the complete form (see Sect. 1.8.1) of this function. There is one-to-one correspondence between arcs of a graph and venjunctions of complete form. So in Fig. 1.3, mentioned correspondence for unity arcs is validated by the following formulae:

$$z = (x \angle y) \vee (x \angle \overline{y}) \vee (\overline{y} \angle x) , \tag{1.30.1}$$

$$z = (x \angle y) \vee (\overline{x} \angle y) \vee (x \angle \overline{y}) \vee (\overline{x} \angle \overline{y}) \vee (\overline{y} \angle \overline{x}) . \tag{1.30.2}$$

1.10 Venjunctive Properties (Basic Formulae)

Being non-Boolean, developed venjunctive operations use its own mathematical apparatus, different from usually applied to binary logic and digital devices as well.

Historical background

Theoretical grounds of Boolean algebra have been basically completed in the 1950s. Since then Boolean logic incorporates the following mathematical instruments.

- Initial operators AND, OR, and NOT named conjunction, disjunction, and negation respectively.
- Logical operator NOR named Peirce arrow [3] and also known as Webb-operation.
- Logical operator NAND named Sheffer stroke [4].
- Venn diagram [5] – a graphical way to represent logical relations between finite sets and binary connectives.
- Disjunctive and conjunctive normal forms, Zhegalkin polynomial [6] – analytical expressions for representation of Boolean operations.
- Truth table – a graphical form for representation of Boolean functions
- Veitch diagram [7] and refined Karnaugh map [8] – a tabular method to simplify Boolean expressions
- Relations between Boolean expressions: commutativity, associativity, distributivity, absorption, De Morgan laws [9], Blake-Poretsky laws [10, 11] etc.

1.10.1 Relations between Operations of Conjunction and Venjunction

Expansion of conjunction into two mirror venjunctions:

$$x \wedge y = (x \angle y) \vee (y \angle x). \tag{1.31}$$

Absorption of conjunction by venjunction:

$$(x \wedge y) \wedge (x \angle y) = (x \angle y). \tag{1.32}$$

Absorption of venjunction by conjunction:

$$(x \wedge y) \vee (x \angle y) = (x \wedge y). \tag{1.33}$$

1.10.2 Inversion (Negation) of Venjunction

In contrast to De Morgan's law, logical negation of venjunction obeys the following rules:

$$\overline{(x \angle y)} = \overline{x} \vee \overline{y} \vee (y \angle x), \tag{1.34.1}$$

$$\overline{(x \angle y)} = \overline{(x \wedge y)} \vee (y \angle x) \, . \tag{1.34.2}$$

Negation of negation, or double negation:

$$\overline{\overline{(x \angle y)}} = (x \angle y) \, . \tag{1.35}$$

1.10.3 Operations with Mirror Venjunctions

Disjunction rules:

$$(x \angle y) \vee (y \angle x) = x \wedge y \, , \tag{1.36.1}$$

$$(x \angle y) \vee \overline{(y \angle x)} = \overline{(y \angle x)} \, , \tag{1.36.2}$$

$$\overline{(x \angle y)} \vee \overline{(y \angle x)} = 1 \, . \tag{1.36.3}$$

Conjunction rules:

$$(x \angle y) \wedge (y \angle x) = 0 \, , \tag{1.37.1}$$

$$(x \angle y) \wedge \overline{(y \angle x)} = x \angle y \, , \tag{1.37.2}$$

$$\overline{(x \angle y)} \wedge \overline{(y \angle x)} = \overline{x \wedge y} \, . \tag{1.37.3}$$

1.10.4 Commutativity and Associativity

Commutativity and associativity rules are not inherent to venjunction:

$$(x \angle y) \neq (y \angle x) \, , \tag{1.38.1}$$

$$x \angle (y \angle z) \neq (x \angle y) \angle z \, . \tag{1.38.2}$$

Along with inequalities, the following formulae are valid:

$$x \angle (y \angle z) = (x \angle y) \wedge (y \angle z) \, , \tag{1.39.1}$$

$$(x \angle y) \angle z = x \angle (y \wedge z) \, , \tag{1.39.2}$$

$$(x \angle y) \angle y = x \angle y \, . \tag{1.39.3}$$

1.10.5 Distributivity and Idempotency

Mainly, distributivity is not inherent to venjunction:

$$x \angle (y \vee z) \neq (x \angle y) \vee (x \angle z) \, , \tag{1.40.1}$$

$$(x \vee y) \angle z \neq (x \angle z) \vee (y \angle z) \, , \tag{1.40.2}$$

$$x \wedge (y \angle z) \neq (x \wedge y) \angle (x \wedge z) \, , \tag{1.40.3}$$

$$x \vee (y \angle z) \neq (x \vee y) \angle (x \vee z) . \tag{1.40.4}$$

As an exception, at least one venjunctive relation conforms to the rule of distribution:

$$x \angle (y \wedge z) = (x \angle y) \wedge (x \angle z) . \tag{1.41}$$

However in the case of mirror components this expression does not obey distributivity:

$$(x \wedge y) \angle z \neq (x \angle z) \wedge (y \angle z) . \tag{1.42}$$

Idempotency is not inherent to venjunction:

$$(x \angle x) \neq x . \tag{1.43}$$

1.10.6 Absorptions

Established above (see Sect. 1.6.4) equality, namely $x \angle y = x \wedge y \wedge \lambda$, allows to form the rules of absorption as follows:

$$x \wedge (x \angle y) = (x \angle y) , \tag{1.44.1}$$

$$y \wedge (x \angle y) = (x \angle y , \tag{1.44.2}$$

$$x \vee (x \angle y) = x , \tag{1.44.3}$$

$$y \vee (x \angle y) = y . \tag{1.44.4}$$

In other cases, including Blake-Poretsky laws, absorption does not happen, and the following inequalities take place:

$$x \angle (x \wedge y) \neq x \wedge y , \tag{1.45.1}$$

$$(x \wedge y) \angle y \neq x \wedge y , \tag{1.45.2}$$

$$x \angle (x \vee y) \neq x , \tag{1.45.3}$$

$$(x \vee y) \angle y \neq y . \tag{1.45.4}$$

$$x \vee (\overline{x} \angle y) \neq x \vee y , \tag{1.45.5}$$

$$x \angle (\overline{x} \vee y) \neq x \angle y . \tag{1.45.6}$$

1.10.7 Rules of Zeroing

In the given context zeroing rules coincide with the similar rules for conjunctions as can be seen from the formula:

$$(x \angle \overline{x}) = (\overline{x} \angle x) = (x \angle 0) = (0 \angle x) = 0 . \tag{1.46}$$

1.10.8   Venjunction with Logical Unity

In the case of venjunction absorptions, which are typical for conjunctions, do not happen. The inequalities take place:

$$x \angle 1 \neq x , \tag{1.47.1}$$

$$1 \angle x \neq x . \tag{1.47.2}$$

1.11   Venjunctive Representation of Indeterminacy

1.11.1   Logical Indeterminacy

If a logic function after and as a result of setting of known values of the variables does not certainly reduce either to zero or unity, such a function is considered to be not determined, that is indeterminate function. Indeterminacy generates ambiguousness in the form of not uniquely defined value $J \in \{0, 1\}$. This uncertain situation can be equally decided in favor of $J = 1$ as well as in favor of $J = 0$.

Indeterminacy of venjunction $z = x \angle y$ is caused by the factor of switching of variables. In this context it is believed that the current values of arguments are known. Provided that unity values $x = y = 1$ take place, the corresponding switching function can not produce a one-valued solution. In contrast, the value $x = 0$ or $y = 0$ ensures a certain value $z = 0$. Other value $z = 1$ is produced at the unity set $[x\ y] = [11]$ if and only if this set have been generated by the switching $x = 0/1$ on the background $y = 1$. In other case of mirror switching $y = 0/1$ on the background $x = 1$, a value $z = 0$ takes place. In the case, when a switching is not determined properly, an appropriative venjunction takes a form of the following operation:

$$(x \angle y) = (1 \angle 1) . \tag{1.48}$$

Thus, as concerning the term of indeterminacy the following laws are postulated:

- if $z \neq 1$ and $z \neq 0$, then $z = J$;
- if $z = 1 \angle 1$, then $z = J$.

The above concept is corroborated and can be proved on the following formal grounds. Earlier (see Sect. 1.6.4) it was defined that a binary variable λ in the formula $z = (x \wedge y) \wedge \lambda$ changes the value exclusively at the moment of switchings $x = 0/1$ and $y = 0/1$. Therefore, if the switchings themselves are only partly determined as $x = J/1$ and $y = J/1$, existing indeterminacy is extended to auxiliary variable, and $\lambda = J$. Thus:

$$z = (x \wedge y) \wedge J . \tag{1.49}$$

As expected in the case of unity values $x = 1$ and $y = 1$, logical indeterminacy is really formed: $z = J$.

1.11.2 Criterion for Indeterminacy

There is a rule, which is valid in law only in reference to indeterminate values. This rule serves as a sort of distinguisher for J in contrast with 1 and 0.

Rule

Logical negation of the indeterminacy is indeterminacy: $\overline{J} = J$.

Assumption

Operation $1 \angle 1$ obeys the rule: $\overline{J} = J$.

Statement

Venjunction of two logical unities is indeterminacy: $1 \angle 1 = J$.

Proof

$$\overline{J} = (\overline{\overline{1 \angle 1}} = \overline{1} \vee \overline{1} \vee (1 \angle 1) = 0 \vee 0 \vee (1 \angle 1) = (1 \angle 1) = J \,. \tag{1.50}$$

References

[1] Vasyukevich, V.: Whenjunction as a logic/dynamic operation. Definition, implementation and applications. Automatic Control and Computer Sciences 18(6), 68–74 (1984)

[2] Vasyukevich, V.: Asynchronous sequences decoding. ACCS Journal 41(2), 93–99 (2007)

[3] Peirce, C.: On the Algebra of Logic. American Journal of Mathematics 3, 15–57 (1880)

[4] Sheffer, H.: A set of five independent postulates for Boolean algebras, with application to logical constants. Transactions of the American Mathematical Society 14, 481–488 (1913)

[5] Venn, J.: On the Diagrammatic and Mechanical Representation of Propositions and Reasonings. Philosophical Magazine and Journal of Science, Ser. 5 10(59) (1880)

[6] Zhegalkin, I.: On the Technique of Calculating Propositions in Symbolic Logic. Mathematical Journal 34(1), 9–28 (1927) (in Russian)

[7] Veitch, E.: A Chart Method for Simplifying Truth Functions. Transactions of the 1952 ACM Annual Meeting, 127–133 (1952)

[8] Karnaugh, M.: The Map Method for Synthesis of Combinational Logic Circuits. Transactions of the American Institute of Electrical Engineers, P. I 72(9), 593–599 (1953)

[9] De Morgan, A.: Formal Logic; or, The Calculus of Inference, Necessary and Probable. Taylor and Walton (1847)

[10] Blake, A.: Canonical expressions in Boolean algebra. The Journal of Symbolic Logic 3(2) (1938)

[11] Poretsky, P.: On methods of solution of logical equalities and on inverse method of mathematical logic. Collected Reports of Meetings of Physical and Mathematical Sciences Section of Naturalists' Society at Kazan University 2 (1984) (in Russian)

Chapter 2
Venjunctive Expressions

Abstract. The second chapter is assigned to demonstrate analytical possibilities of venjunction as applied to memory devices and functions. Behavior of bistable cell and most significant trigger-type circuits are analyzed in detail. Logical circuits of venjunctor are proposed as devices intended for realizing function of venjunction. By the example of double venjunctor it is clearly demonstrated the way of the corresponding logical formula restoration. It is also shown that unique analytical abilities of venjunction are extended to devices with logically indeterminate function.

2.1 Bistable Cell

Bistable cell (BC) is the simplest memory element in asynchronous (venjunctive) logic. This cell is considered as a fundamental element for constructing and realizing the operation of venjunction itself as well as of venjunctive functions on its basis.

A structural diagram of bistable cell is presented in Fig. 2.1. It consists of two two-input logical elements of NAND, whose outputs are cross-connected with each other inputs.

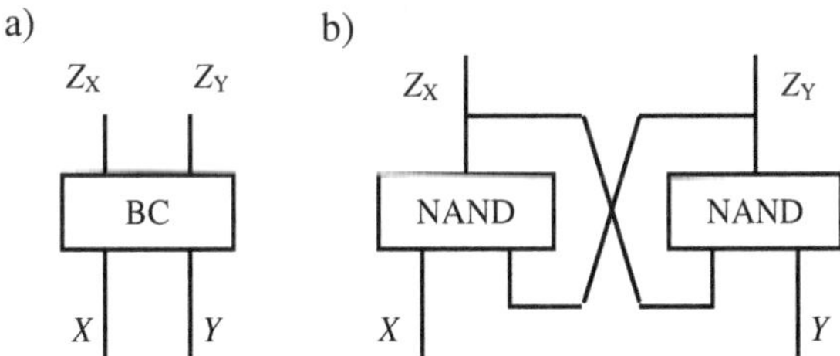

Fig. 2.1 Bistable cell built with NAND-elements: a) BC element; b) logical circuit

Binary cell is symmetrical by its own structure. There are two inputs designated as X and Y, and two outputs Z_X and Z_Y respectively. Switchings of input signals are transformed by logical NAND-elements into output switchings. Thus, the logic of bistable cell operating is defined. Corresponding logical relationships are displayed in the form of switching tables.

V. Vasyukevich: Asynchronous Operators of Sequential Logic, LNEE 101, pp. 25–41.
springerlink.com   © Springer-Verlag Berlin Heidelberg 2011

2.1.1 Functions of Bistable Cell

Logical operations inherent to bistable cell (Fig. 2.1) are represented by a table of switching. Table 2.1 show what kind of binary switchings appear on the outputs of bistable cell in response to the switchings of input signals. For example, appearance $Z_X = J/1$ in response to $X = 1/0$ and $Y = 1/1$ means that owing to switching $X = 1/0$ on the background $Y = 1$ bistable cell is transferred to unity state of the output Z_X. Sign J symbolizes dependency of the output signal from its initial, and therefore unknown, value.

Table 2.1 Switchings of signals of bistable cell (Fig. 2.1): a) output Z_X; b) output Z_Y.

a)

		Y		
	1/1	0/1	1/0	0/0
1/1	J/J	0/0	J/0	0/0
0/1	1/1	-	-	1/0
1/0	J/1	-	-	0/1
0/0	1/1	1/1	1/1	1/1

(X labels the rows)

b)

		Y		
	1/1	0/1	1/0	0/0
1/1	J/J	1/1	J/1	1/1
0/1	0/0	-	-	1/1
1/0	J/0	-	-	1/1
0/0	0/0	1/0	0/1	1/1

(X labels the rows)

According to Sect. 1.4.3, venjunctive function is defined at such values that complete switchings of output signal in tables (Table 2.1). In view of the given circumstance, the examined bistable cell responds to input switchings in compliance with the data of tables in Table 2.2. These tables represent the following functional dependences:

$$Z_X = F_X (X,Y), \tag{2.1.1}$$

$$Z_Y = F_Y (X,Y), \tag{2.1.2}$$

where $X \in \{1, 0/1, 1/0, 0\}$, $Y \in \{1, 0/1, 1/0, 0\}$, $Z \in \{1, 0\}$.

Table 2.2 Bistable cell (Fig. 2.1) functioning: a) output Z_X; b) output Z_Y.

a)

		Y		
	1	0/1	1/0	0
1	-	0	0	-
0/1	1	-	-	0
1/0	1	-	-	1
0	-	1	1	-

(X labels the rows)

b)

		Y		
	1	0/1	1/0	0
1	-	1	1	-
0/1	0	-	-	1
1/0	0	-	-	1
0	-	0	1	-

(X labels the rows)

Transforming tabulated switchings into venjunctions, and applying these venjunctions to the given above dependences, the following logical expressions are formed:

$$Z_X = (X \angle Y) \vee (\bar{X} \angle Y) \vee (Y \angle \bar{X}) \vee (\bar{Y} \angle \bar{X}) \vee (\bar{X} \angle \bar{Y}), \tag{2.2.1}$$

$$Z_Y = (Y \angle X) \vee (\bar{Y} \angle X) \vee (X \angle \bar{Y}) \vee (\bar{X} \angle \bar{Y}) \vee (\bar{Y} \angle \bar{X}). \tag{2.2.2}$$

These expressions are output functions presented in their venjunctive complete form. They combine all input venjunctions, unity values of which cause the outputs of $Z_X = 1$ and $Z_Y = 1$. After minimizations (see Sect. 1.10.3, Eq. 1.36.1) on the basis of equalities:

$$(\bar{X} \angle Y) \vee (Y \angle \bar{X}) = \bar{X} \wedge Y, \tag{2.3.1}$$

$$(\overline{X} \angle \bar{Y}) \vee (\overline{Y} \angle \bar{X}) = \bar{X} \wedge \bar{Y}, \tag{2.3.2}$$

$$\bar{X} \wedge Y \vee \bar{X} \wedge \bar{Y} = \bar{X}, \tag{2.3.3}$$

output functions are reduced to the compact forms as follows:

$$Z_X = \bar{X} \vee (X \angle Y), \tag{2.4.1}$$

$$Z_Y = \bar{Y} \vee (Y \angle X). \tag{2.4.2}$$

Thus, unity value of output Z_X (Z_Y) takes place at zero signal of input X (Y), or at the switching $X = 0/1$ ($Y = 0/1$) on the background $Y = 1$ ($X = 1$). Specifics of formulae allow their otherwise reading, for example referring to Z_X:

-   if signal $X = 0$ occurs, function takes unity value and retains it, even when the value of X switches, but only until a signal $Y = 1$ is unchangeable, that is until the moment of switching $Y = 1/0$.

For $Z_X = 0$ (Table 2.2), a compact expression and its complete form are presented as follows:

$$\bar{Z}_X = X \wedge \bar{Y} \vee (Y \angle X), \tag{2.5.1}$$

$$\bar{Z}_X = (X \angle \bar{Y}) \vee (\bar{Y} \angle X) \vee (Y \angle X). \tag{2.5.2}$$

Similar expressions for zero value $Z_Y = 0$ are given by the formulae:

$$\bar{Z}_Y = \bar{X} \wedge Y \vee (X \angle Y), \tag{2.6.1}$$

$$\bar{Z}_Y = (\bar{X} \angle Y) \vee (Y \angle \bar{X}) \vee (X \angle Y). \tag{2.6.2}$$

2.1.2 Bistable Cell with NOR Elements

Bistable cell based on the logical elements of NOR is presented by its structural diagram in Fig. 2.2.

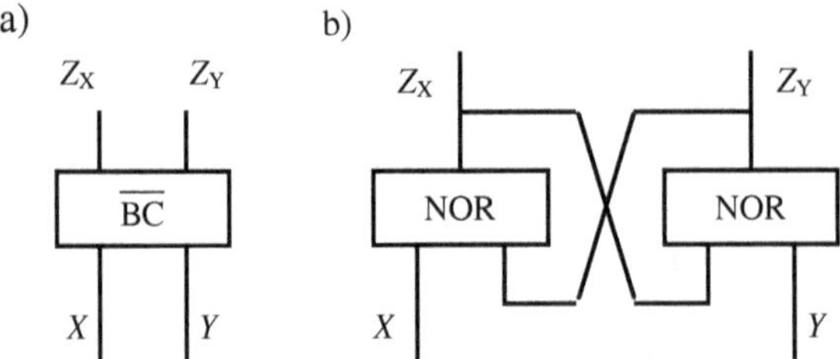

Fig. 2.2 Bistable cell built with NOR-elements: a) $\overline{BC}$ element (inverted BC); b) logical circuit

Output signals Z_X and Z_Y depending on the input switchings are displayed in Table 2.3. From comparing the data in Table 2.3 with the corresponding data in Table 2.2 follows that after replacement of logical NAND-elements (Fig. 2.1) with NOR-elements (Fig. 2.2) functioning of bistable cell changes in a certain way.

Table 2.3 Bistable cell (Fig. 2.2) functioning: a) output Z_X; b) output Z_Y.

a)

X		Y		
	1	0/1	1/0	0
1	-	0	0	-
0/1	0	-	-	0
1/0	1	-	-	0
0	-	1	1	-

b)

X		Y		
	1	0/1	1/0	0
1	-	0	1	-
0/1	0	-	-	1
1/0	0	-	-	1
0	-	0	0	-

Former logic operations and structure of formulae remain valid, but input and output signals are subject to negation. Thus, functions assume the form of dependences:

$$\overline{Z}_X = F_X(\overline{X},\overline{Y}), \tag{2.7.1}$$

$$\overline{Z}_Y = F_Y(\overline{X},\overline{Y}). \tag{2.7.2}$$

The corresponding changes have an effect on venjunctive expressions. Logic of bistable cell (Fig. 2.2) functioning is defined by formulae:

$$\overline{Z}_X = X \vee (\overline{X} \angle \overline{Y}), \tag{2.8.1}$$

$$\overline{Z}_Y = Y \vee (\overline{Y} \angle \overline{X}), \tag{2.8.2}$$

$$Z_X = \overline{X} \wedge Y \vee (\overline{Y} \angle \overline{X}), \tag{2.8.3}$$

$$Z_Y = X \wedge \overline{Y} \vee (\overline{X} \angle \overline{Y}). \tag{2.8.4}$$

2.2 Triggers

A trigger is assumed to be a common name for all kinds of binary devices (including flip-flops, latches, bistable multivibrators, etc.) with two stable output states. Each state, being set, remains stable until switching signal appears. Logical circuits for triggers of various types are constructed on the basis of bistable cell.

2.2.1 SR Flip-Flop

The logic of trigger functioning is given in Table 2.4.

Table 2.4 SR flip-flop functioning.

		R			
		1	0/1	1/0	0
	1	-	F	1	-
S	0/1	F	-	-	1
	1/0	0	-	-	1
	0	-	0	0	-

Symbols S and R denote the input signals. Output signal of the device is marked as Q. The S input is intended for setting the trigger in its unity state, that means logical unity appearance at the Q output: $Q = 1$. A similar unity signal at R input produces a switching of trigger into its zero state, that is $Q = 0$.

The special feature of SR flip-flop is in its input regulation, due to that a coincidence of signals $S = 1$ and $R = 1$ is not allowed. In other words, it is forbidden to perform unity and zero setting actions simultaneously. This circumstance is symbolized by character "F".

Taking into account Table 2.4, it is assuming that only permitted switchings are considered, the function of SR flip-flop appears in the following venjunctive complete form:

$$Q = (S \angle \overline{R}) \vee (\overline{R} \angle S) \vee (\overline{S} \angle \overline{R}) . \tag{2.9}$$

In compact form this function is expressed by the formula:

$$Q = S \wedge \overline{R} \vee (\overline{S} \angle \overline{R}) . \tag{2.10}$$

Inversion of the given formula gives an expression:

$$\overline{Q} = ((\overline{S} \vee R) \wedge (S \vee R \vee \overline{R} \angle \overline{S})) = R \vee \overline{R} \angle \overline{S} , \tag{2.11}$$

that does not agree with the data in Table 2.4, because aside from allowed input values, a forbidden binary combination $[S\ R] = [11]$ is also able to perform zero setting $Q = 0$. This disadvantage (irrelevance) is avoided analytically by updating above mentioned expression as follows:

$$\overline{Q} = \overline{S} \wedge R \vee (\overline{R} \angle \overline{S}). \tag{2.12}$$

For SR flip-flop implementation two solutions at least are applicable.

1. Bistable cell based on NAND-elements (Fig. 2.1), when inputs and outputs are associated with trigger signals in accordance with the following equalities:

   – $X = \overline{S}$, $Y = \overline{R}$, $Z_X = Q$, $Z_Y = \overline{Q}$.

2. Bistable cell based on NOR-elements (Fig. 2.2), when inputs and outputs are associated with trigger signals in accordance with the following equalities:

   – $X = S$, $Y = R$, $Z_X = \overline{Q}$, $Z_Y = Q$.

Remark

In spite of the fact that SR flip-flop is based on an element of asynchronous logic, the trigger itself can be considered as asynchronous device only conditionally. This is because one of the input binary sets is forbidden, that does not perfectly match with principles of asynchronous implementations. Although, on the other hand, exactly owing to the given restriction, SR flip-flop can be named a trigger as such.

2.2.2   JK Flip-Flop [1]

One of the possible solutions for constructing an asynchronous trigger is that the SR flip-flop functionality can be expanded at the expense of permission to use previously forbidden input binary set $[S\ R] = [11]$. In other words, it is necessary to specify the values marked "F" in Table 2.4 by replacing these marks with logical symbols 0 and 1. Thus, the function of a trigger becomes completely determined, and not caused by restrictions unacceptable for the asynchronous logic.

To ensure that updated SR flip-flop is transformed into JK flip-flop the following changes in Table 2.4 are required. Switching $S = 0/1$ on the background $R = 1$ should be applied for $Q = 1$ setting, and switching $R = 0/1$ on the background $S = 1$ for $Q = 0$ setting. Furthermore, S input should be replaced with J, and R input – with K. As a result, the function displayed in Table 2.5, as well as the JK trigger, is formed.

---

[1] Not to be confused with J, a symbol of logical indeterminacy.

Table 2.5 JK flip-flop functioning.

		K			
		1	0/1	1/0	0
	1	-	0	1	-
J	0/1	1	-	-	1
	1/0	0	-	-	1
	0	-	0	0	-

According to the table, logic of **JK** flip-flop functioning is defined by the following venjunctive expressions:

$$Q = (J \angle K) \vee (\overline{K} \angle J) \vee (J \angle \overline{K}) \vee (\overline{J} \angle \overline{K}), \qquad (2.13.1)$$

$$Q = J \wedge \overline{K} \vee (J \angle K) \vee (\overline{J} \angle \overline{K}). \qquad (2.13.2)$$

Another analytical representation of the trigger is given by the formula:

$$Q = (J \wedge \overline{K} \vee J \angle K) \vee \overline{(J \wedge \overline{K} \vee J \angle K)} \angle \overline{(\overline{J} \wedge K \vee K \angle J)}, \qquad (2.14)$$

containing logical expressions $(J \wedge \overline{K} \vee J \angle K)$ and $(\overline{J} \wedge K \vee K \angle J)$, which are assigned for output unity and zero settings respectively.

Classical implementation of JK flip-flop is based on the feedback usage with the output signal participation. As a result, the function takes the form of functional dependence $Q = F(J, K, Q)$, which is specified by the formula:

$$Q = (J \wedge \overline{Q}) \vee \overline{(J \wedge \overline{Q})} \angle \overline{(K \wedge Q)}. \qquad (2.15)$$

From an asynchronous logic standpoint, the obtained function is not wholly correct, because there are input switchings with indefinite output after-effects. While $J = 1$ setting at the time of $K = 1$ and $Q = 0$ takes place, the output signal changes because of the switching $Q = 0/1$. Under this switching the following changes occur: $(J \wedge \overline{Q}) = 1/0$ and $(K \wedge Q) = 0/1$. If applied to venjunction $\overline{(J \wedge \overline{Q})} \angle \overline{(K \wedge Q)}$, the occurred changes do not ensure an expected switching $\overline{(J \wedge \overline{Q})} = 0/1$ on the background $\overline{(K \wedge Q)} = 1$, while it is necessary for output signal to be kept in its unity state. In these circumstances a reverse switching $Q = 1/0$ is admissible as well. So, it is possible for output signal to return to the initial zero state that symbolizes an unsuccessful effort to perform the unity value setting: $Q = 1$.

In actual practice, a problem of functional indeterminacy is considered and solved as a problem of race hazard – dangerous race of signals. To avoid these hazards various schematic and technical methods are used. Usually, such time delays and such ways for signals passing are designed, that the "right" signal always turns out to be a winner.

In principle, existing race problem can be solved even at a function level. It is sufficient to block undesirable signals, and in this way to exclude indeterminacy. As to JK trigger, a possible indeterminations are caused by switchings $Q = 0/1$ and $Q = 1/0$ on the background $[J\ K] = [11]$. This problem can be solved analytically, for example, by means of the following function:

$$Q = (J \angle \overline{Q}) \vee \overline{(J \angle \overline{\overline{Q}})} \angle (\overline{K \angle Q}) , \qquad (2.16)$$

where venjunctions $J \angle \overline{Q}$ and $K \angle Q$ are involved for performing unity and zero settings respectively. Corresponding switchings remove the problem of hazards. Due to the given updates a JK flip-flop is free from hazardous functional race.

Remark
Generally JK flip-flop is realized as synchronous device with additional clock signal. Therefore, it will be more correct to represent updated trigger as an asynchronous module of JK trigger.

2.2.3 T Flip-Flop (Toggle Trigger)

There are two types of T flip-flops: static and dynamic. They differ from each other by the manner of acting of the clocking procedure. Static trigger is toggled responding to the clock signal identified by its value, and dynamic trigger responds to the switching of the signal. The logic of static toggle trigger functioning is presented in Table 2.6.

Table 2.6 Static T flip-flop functioning.

		Q	
		1	0
T	1	0	1
	0	1	0

In static mode T flip-flop realizes the following function:

$$Q = T \wedge \overline{Q} \vee \overline{(T \wedge \overline{\overline{Q}})} \angle (\overline{T \wedge Q}) . \qquad (2.17)$$

For this function (as well as in the case of JK flip-flop), a race problem is characteristic. Switching $T = 0/1$ on the background $Q = 0$ produces output switching $Q = 0/1$, which provokes race process prolonged up to the moment of $T = 1/0$. At this time a stable output state is formed, and its signal can be set to zero as well as to unity value. In the first event after output setting $Q = 1$ invalid switching $Q = 1/0$ takes place. In other case trigger maintains the unity state, and so performs its own toggle function.

In an effort to provide a valid functioning for the T flip-flop, race hazards must be blocked, for example, by the clock pulse reduction. Combining mentioned above and other schematic and technical methods, it is possible to avoid the hazardous switchings $Q = 0/1$ and $Q = 1/0$ on the background $T = 1$ completely.

Dynamic T flip-flop in contrast to static trigger is free from dangerous problem caused by raced signals. All solutions related with blocking of hazardous switchings are avoided not during the process of construction or realization, but beforehand – when logical function is produced. Considering that input signal always overtakes returned output signal (in practice such assumption is quite justified), dynamic trigger can be implemented on the base of the following venjunctive function:

$$Q = T \angle \overline{Q} \vee \overline{(T \angle \overline{Q})} \angle (\overline{T} \angle Q). \qquad (2.18)$$

Obtained toggle trigger is dynamical because of its output settings, that are performed not during the time of unity signal $T = 1$, but at the moment of switching $T = 0/1$. All possible input switchings therewith are allowed and each of them causes a valid hazard-free transition to the next state. These states are presented by trigger output values in Table 2.7.

Table 2.7 Dynamic T flip-flop functioning.

		Q			
		1	0/1	1/0	0
	1	-	1	0	-
T	0/1	0	-	-	1
	1/0	1	-	-	0
	0	-	1	0	-

Remark
Dynamic toggle trigger is able to function without any restrictions in relation to input signals, and therefore dynamic T flip-flop belongs to asynchronous logic.

2.2.4 Clocked D-Latch

In analogy with a toggle trigger, clocked D-latches are also subdivided into two types. Static and dynamic triggers differ from one another by their functional possibilities. Corresponding Table 2.8 clearly demonstrate how mentioned differences are presented on the outputs of compared types of triggers.

D-latch is clocked by T signal. Aside from the well-known constant values (0, 1), and used above indeterminate value (J), behavior of the given circuit is characterized by symbol J' with apostrophe. In contrast with J, whose value depends on switching that takes place, a value of J' depends on a chain, that is, a sequence of switchings. Symbol J' characterizes so called secondary indeterminacy, when a signal is not certainly defined, because it depends on another signal or on the previous value of the same signal. In the context of D-latch, updated symbol denotes, that a previous output signal is repeated; it is stored according to pseudoswitchings $Q = 0/0$ or $Q = 1/1$. In any case J' follows J or J'.

Table 2.8 Clocked D-latch functioning: a) static type; b) dynamic type.

a)

		\(D\)			
		1	0/1	1/0	0
	1	-	1	0	-
T	0/1	1	-	-	0
	1/0	1	-	-	0
	0	-	J'	J'	-

b)

		\(D\)			
		1	0/1	1/0	0
	1	-	J'	J'	-
T	0/1	1	-	-	0
	1/0	J'	-	-	J'
	0	-	J'	J'	-

Analytically, functions of static and dynamic triggers are expressed by the following formulae:

$$Q = T \wedge D \vee \overline{(T \wedge D)} \angle \overline{(\overline{T \wedge D})}, \qquad (2.19.1)$$

$$Q = T \angle D \vee (\overline{(T \angle D)} \angle \overline{(\overline{T \angle D})}). \qquad (2.19.2)$$

Static D-latch is able to change its output signal repeatedly during a clock period when $T = 1$. This circumstance produces independency as a factor of trigger behavior, because it is not known, what a state will be set at the moment of clock signal reverse switching $T = 1/0$. To avoid confusion, it is usually recommended to shorten clock pulse. However, even a short clock signal does not eliminate but only reduces the hazard of trigger not proper functioning. To solve this problem completely, it is necessary to block dangerous switchings $D = 0/1$ and $D = 1/0$ on the background $T = 1$, or to assign output values for these switchings. The last solution is basically used for dynamic type triggers constructing purposes.

Setting actions for dynamic D-latch are performed by venjunctions instead of conjunctions related to static latch. As a result of this replacement, trigger becomes free from invalid consequences caused by repeated switchings. Changes of output state are exclusively controlled by switching $T = 0/1$, under impact of which trigger reaches a stable condition. Signal of D input remains blocked until the next switching $T = 0/1$ appears.

When clock pulse is regulated in time, static D-latch functions as a synchronized logical device. As to dynamic trigger, its belonging to asynchronous logic is undoubted.

Note

Results of D-latch switchings (Table 2.8) are partly not determined. There are switchings in response to which, function Q reacts ambiguously (J' symbol). It means that memory depth of trigger device exceeds the level which can be provided and maintained by the elementary venjunctions. Due to this reason the function of D-latch cannot be expanded into components of venjunctive complete form (see Sect. 1.8.1).

Historical background

In 1918 the first flip-flop (Eccles–Jordan trigger [1]) was patented, and later in 1941 the first computer was invented [2]. At the same time it was suggested to implement Boolean algebra operations by means of relay circuits [3, 4]. The corresponding researches and also cybernetics [5, 6] as a novel science direction largely stimulated the following intensive development of the circuit engineering. It concerned not only combinational circuits but sequential circuits too, because all types of memory elements were already defined [7].

2.3 Venjunctor

Venjunctor [8] is a two-input logical device intended for realization of venjunctive functions, such as in [9] for R(t)-trigger or for identifier in [10]. Being a logical element the corresponding circuit is build on the basis of bistable cell. Functional properties of this cell allow constructing a venjunctor using different methods.

2.3.1 Expressions and Circuits of Venjunctor

If a bistable cell (Fig. 2.1) is applied, it is suitable to deduce a venjunction $X \angle Y$ by the following transformations:

$$Z = (X \wedge Z_X = X \wedge (\overline{X} \vee X \angle Y)) = X \angle Y , \qquad (2.20.1)$$

$$Z = (X \wedge \overline{Z}_Y = X \wedge (\overline{X} \wedge Y \vee X \angle Y)) = X \angle Y , \qquad (2.20.2)$$

$$Z = (X \wedge Y \wedge Z_X) = X \angle Y , \qquad (2.20.3)$$

$$Z = (X \wedge Y \wedge \overline{Z}_Y) = X \angle Y . \qquad (2.20.4)$$

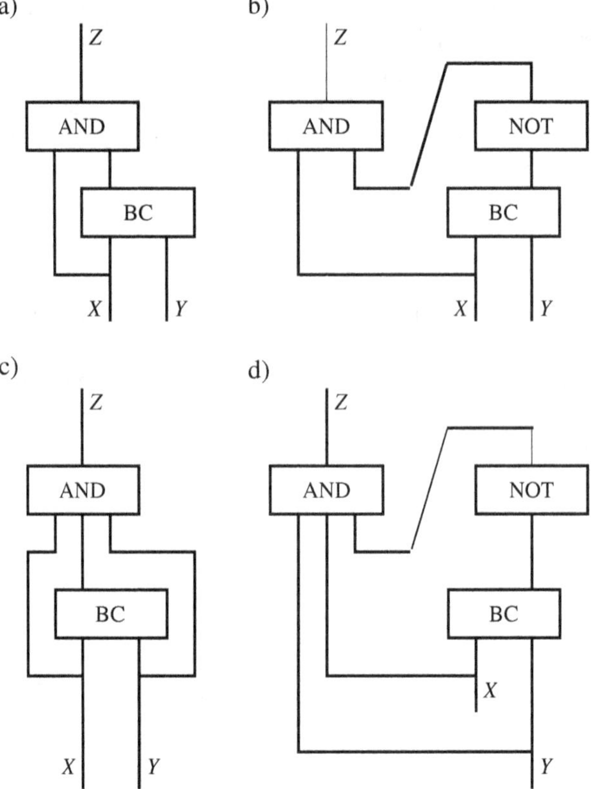

Fig. 2.3 Logical circuits of venjunctor, designed by:
Eq. 2.20.1; b) Eq. 2.20.2; c) Eq. 2.20.3; d) Eq 2.20.4

The given four formulae serve as a reference for constructing logical circuits (Fig. 2.3). Each of these circuits performs the operation of venjunction: $Z = X \angle Y$. BC – bistable cell (see Sect. 2.1.1, Fig. 2.1) is a basic logical element for venjunctor implementation.

From a structural redundancy and signals race standpoint, it can be determined that the logical circuit shown in Fig. 2.3b realizes functionality of venjunctor most efficiently.

Venjunctor is a two-input logical device intended for implementing venjunctive function. Venjunctor as bistable V-element of asynchronous logic is displayed in Fig. 2.4.

Taking into account that venjunction is asymmetrical operation V-element is displayed in two forms. Output Z is displaced in the direction of either X input (Fig. 2.4a) or Y input (Fig. 2.4b). In this way it is denoted that inputs of venjunctor are not graphically equal in their relations with the output. Fig. 2.4a presents venjunctor, which is intended to realize a switching $X = 0/1$ on the background $Y = 1$. Accordingly, X input is assumed to be active in contrast to passive input Y. Otherwise in Fig. 2.4b X input is passive and Y input – active; switching $Y = 0/1$ on the background $X = 1$ is performed.

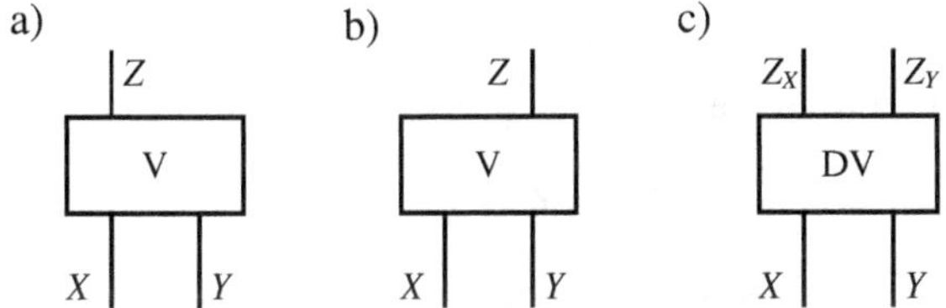

Fig. 2.4 a) Venjunctor with function $Z = X \angle Y$; b) Venjunctor with function $Z = Y \angle X$; c) DV – Double Venjunctor

If venjunctive element has two outputs, it is double venjunctor, marked DV (Fig. 2.4c). Double venjunctor unites two venjunctors so that two mirror venjunctive functions are performed.

$$Z_X = X \angle Y , \tag{2.21.1}$$

$$Z_Y = Y \angle X . \tag{2.21.2}$$

Note

Circuit implementations, that operate using inverted signals, are ensured by the logical $\overline{BC}$-element (Fig. 2.2). In this case the following formulae can be involved:

$$\overline{X} \wedge \overline{Z}_X = (\overline{X} \wedge (X \vee \overline{X} \angle \overline{Y})) = \overline{X} \angle \overline{Y} , \tag{2.22.1}$$

$$\overline{X} \wedge Z_Y = (\overline{X} \wedge (X \wedge \overline{Y} \vee \overline{X} \angle \overline{Y})) = \overline{X} \angle \overline{Y} . \tag{2.22.2}$$

2.3.2 Double Venjunctor

Double venjunctor is presented by its logical circuit in Fig. 2.5. This circuit contains two bistable cells connected consecutively in such a manner that outputs of the first cell are linked with the inputs of another cell. Input signals are applied to both cells. On the circuit outputs, two inventors (NOT-elements) are placed.

According to the logical circuit, dependence of output signal Z_X from input signals X and Y is represented by the following formula:

$$Z_X = \overline{(X \wedge (\overline{X} \vee X \angle Y)) \vee ((X \wedge (\overline{X} \vee X \angle Y)) \angle (Y \wedge (\overline{Y} \vee Y \angle X)))} . \tag{2.23}$$

Taking into account equalities

$$- \quad (X \wedge (\overline{X} \vee X \angle Y)) = X \angle Y \ \text{ and } \ (Y \wedge (\overline{Y} \vee Y \angle X)) = Y \angle X \ ,$$

the given expression is minimized as follows:

$$Z_X = (\overline{\overline{X \angle Y}}) \vee ((X \angle Y) \angle (Y \angle X)) . \tag{2.24}$$

Further logical actions, including zeroing $(X \angle Y) \angle (Y \angle X) = 0$, produce a venjunction as a result of the following operation:

$$Z_X = (\overline{\overline{X \angle Y}}) = X \angle Y . \tag{2.25}$$

Similar transformations related to output signal Z_Y ensure deducibility of another (mirror) venjunction:

$$Z_Y = (\overline{\overline{Y \angle X}}) = Y \angle X . \tag{2.26}$$

Thus, as to double venjunctor, its functional validity is certainly confirmed by the logical circuit in Fig. 2.5.

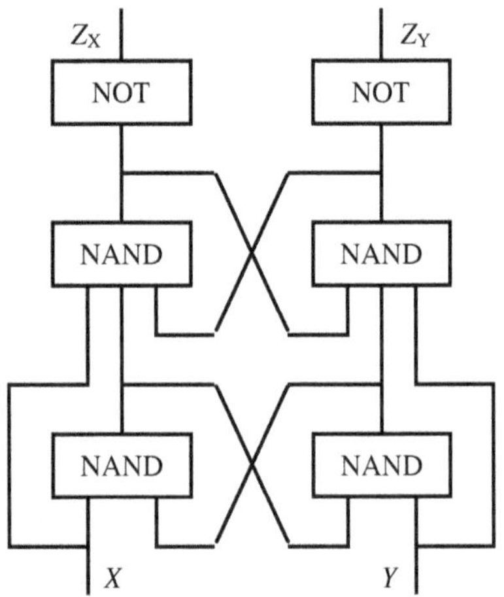

Fig. 2.5 Logical circuit of double venjunctor

2.4 Logical Circuits with "Exotic" Functions

As exotic are named functions, which are practically useless. Their only "justification" is that they exist and can be expressed analytically as well as displayed in forms of tables and logical circuits. Corresponding functions take place in the case of venjunctive operations with a constant value – logical unity.

2.4.1 *Truncated Venjunction with Function* $Z = X \angle 1$

Functioning of logically truncated venjunction is presented in Table 2.9.

Table 2.9 Venjunction $Z = X \angle 1$: a) signals switchings, b) switching function.

a)			Y	b)			Y
			1/1				1
		1/1	J/J			1	J
	X	0/1	0/1		X	0/1	1
		1/0	J/0			1/0	0
		0/0	0/0			0	0

Corresponding realization is made on the basis of venjunctor (Fig. 2.3) keeping in view that the functionality is restricted (truncated) because of the constant value $Y = 1$. In these conditions the resulting logical circuit is displayed in Fig. 2.6.

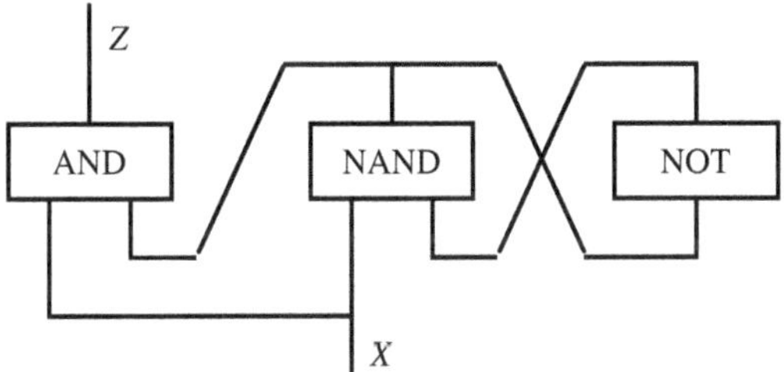

Fig. 2.6 Logical circuit of venjunction $Z = X \angle 1$

2.4.2 *Truncated Venjunction with Function* $Z = 1 \angle Y$

Functioning of logically truncated venjunction is presented in Table 10.

Table 2.10 Venjunction $Z = 1 \angle Y$: a) signals switchings, b) switching function.

a)		Y				b)		Y			
		1/1	0/1	1/0	0/0			1	0/1	1/0	0/0
X	1/1	J/J	0/0	J/0	0/0	X	1	J	0	0	0

Corresponding realization is made on the basis of venjunctor (Fig. 2.3) keeping in view that the functionality is truncated because of the constant value $X = 1$. In these conditions the resulting logical circuit is displayed in Fig. 2.7.

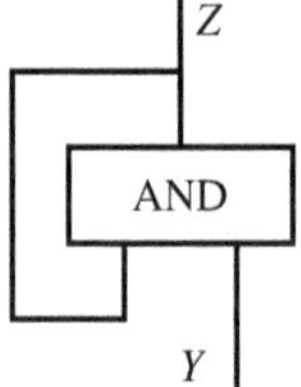

Fig. 2.7 Logical circuit of function $Z = 1 \angle Y$

2.4.3   Indefinite Venjunction with Degenerative Function $Z = 1 \angle 1$

The logic of venjunction functioning is presented in Table 2.11, where it is seen that the function degenerates into indeterminacy.

Table 2.11 Indefinite function $Z = 1 \angle 1$.

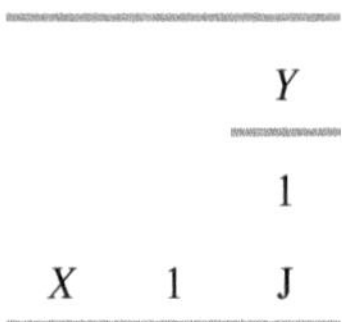

		Y
		1
X	1	J

Corresponding realization is made on the basis of venjunctor (Fig. 2.3) keeping in view that the functionality is restricted because of the constant values $X = 1$ and $Y = 1$. In these conditions the resulting "logical circuit" is displayed in Fig. 2.8.

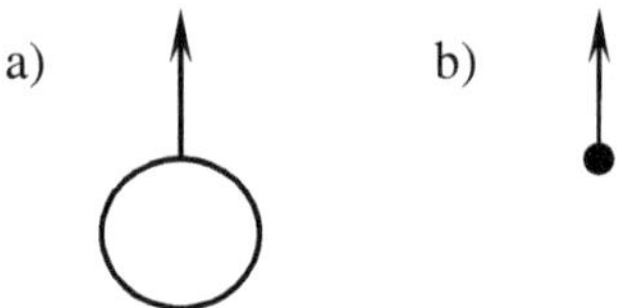

Fig. 2.8 "Circuits" of logical indeterminacy $1 \angle 1$:
a) element without input – dead loop; b) point element – output from nowhere.

Remark

Above mentioned examples are remarkable because of their clarity. It seems to be the simplest way to demonstrate those not trivial possibilities that become available owing to mathematics of venjunctive formulae.

References

[1] Eccles, W., Jordan, F.: A trigger relay utilizing three-electrode thermionic vacuum tubes. The Electrician 83, 298 (1919)

[2] Zuse, K.: Method for Automatic Execution of Calculations with the Aid of Computers. German Pat. App., Z 23 139 IX/42m (1936)

[3] Shannon, C.: A Symbolic Analysis of Relay and Switching Circuits. Trans AIEE 57, 713–723 (1938)

[4] Shestakov, V.: Algebra of Two Poles Schemata (Algebra of A-Schemata). Automatics and Telemechanics (2), 15–24 (1941) (in Russian)

[5] von Neumann, J.: Cybernetics, or Control and Communication in the Animal and Machine. Hermann & Cie (1948)

[6] Wiener, N.: Cybernetics: Or Control and Communication in the Animal and the Machine. MIT Press, Cambridge (1948)

[7] Phister, M.: Logical Design of Digital Computers. Wiley, Chichester (1958)

[8] Vasyukevich, V.: Whenjunction as a logic/dynamic operation. Definition, implementation and applications. Automatic Control and Computer Sciences 18(6), 68–74 (1984)

[9] Vasyukevich, V.: Enumeration of tests for sequence circuits. ACCS Journal 25(5), 78–84 (1991)

[10] Vasyukevich, V.: Monotone sequences of binary data sets and their identification by means of venjunctive functions. ACCS Journal 32(5), 49–56 (1998)

Chapter 3
Sequention

Abstract. The third chapter continues a theme of the first section on a higher level. Not only separate switchings are investigated but their sequences of a certain type. These sequences are named sequentions. Sequention is a more complicated mathematical form, which includes venjunctions as minisequentions. Depending on configuration, sequentions are subdivided into the following types: simple and complicated, correct and incorrect, perfect and imperfect, elementary and composite, compatible and inconsistent. For the elements of composite sequentions, binary relations are established as being derived from relationships existing in ordered sets. The rules are developed for transformation of sequentions by means of splitting and slicing actions. General principles are formulated for performing conjunction, disjunction and venjunction operations on sequentions. On the basis of decomposition of sequentions logical expressions are obtained. These expressions represent sequential functions in conjunctive or venjunctive form. For clarity, logical behavior of sequention is assigned by a graph of special form. It is suggested to estimate logical expressions of sequential type by memory depth and volume of memory. Example of the corresponding calculation is brought forward.

3.1 Sequention as an Ordered Set

Initially defined that sequention is a sequence of Boolean variables. At the same time, sequention may be considered as an ordered set as well as a function of this set.

In the context of ordered set, sequention (*monotone sequence* as prototype in [1]) is a sequence of one-type elements, for example x. Elements of sequention are indexed by numbers of positions; according to those they are sequentially located. Thus, in the sequention:

$$\langle x \rangle = \langle x_1\, x_2\, \dots\, x_n \rangle \tag{3.1}$$

element x_1 occupies the first position, x_2 – the second, and so on until x_n. Number "n" assigns a length of ordered sequence (number of elements), and therefore n-positioned sequention is specified as n-sequention.

The set $\{x_1\, x_2\, \dots\, x_n\}$ does not contain similar, equal one to another elements. In other words sequention should not permit repeated elements.

Every sequention is finite in length, and it incorporates initial (First) element $x_F = x_1$ and final (Last) element $x_L = x_n$ in accordance with $\langle x_1\, x_2\, \dots\, x_n \rangle$.

V. Vasyukevich: Asynchronous Operators of Sequential Logic, LNEE 101, pp. 43–75.
springerlink.com   © Springer-Verlag Berlin Heidelberg 2011

Current elements of sequentions are provided by indexes i and j so that $x_i \in \langle x \rangle$ and $x_j \in \langle x \rangle$, where $(i, j) \in \{1, 2, \dots n\}$. It is meant (if is not especially conditioned) that $x_i \neq x_j$. But if equality of elements is allowed, the corresponding indexes are also equal, that is $i = j$. Otherwise initially assumed principle of uniqueness of elements is violated.

Order of elements in sequention is conditioned by rules, which are intended for regulating placement of elements in common sequence. These rules define relations of elements with each other by means of priority sign "$\prec$" or "$\succ$". Every pair of elements is linked by binary relation $x_i \prec x_j$, according to which element x_i leads, and is ahead of x_j because of $i < j$. This pair can be also characterized by equivalent relationship $x_j \succ x_i$, where element x_j lag behind x_i. From a sequention standpoint, relation $x_i \prec x_j$, as well as $x_j \succ x_i$, means that element x_j follows x_i.

Sequention, being an ordered set, obeys the following transitive assertions:

- if $x_i \prec x_j$, then $x_i \prec x_{j+1}$;
- if $x_i \prec x_j$, then $x_{i-1} \prec x_j$.

If element x_i is ahead of x_j, the same element is also ahead of x_{j+1}, and the previous element x_{i-1} is ahead of x_j.

Consequence

The given binary relations are generalized by the following statement. If all adjacent elements are connected by the binary relation $x_i \prec x_{i+1}$, where $i \in \{1, 2 \dots n-1\}$, the corresponding ordered sequence is a sequention.

Remark on Intersection of Sets

All elements of sequention are arranged in a certain order, and this circumstance must be taken into account when operation of intersection is performed. That is because a nonempty intersection of unordered sets, $\{x \cap y\} = \{z\}$ as example, not in the least exclude the possibility of empty intersection of the corresponding sequentions: $\langle x \rangle \cap \langle y \rangle = \varnothing$. Such a discrepancy might take place when the same elements of $\{z\}$ set are included in sequentions $\langle x \rangle$ and $\langle y \rangle$ so that their orders differ from each other. But if the orders are in agreement, then $\langle x \rangle \cap \langle y \rangle = \langle z \rangle$. Furthermore, if $\{x \cap \{y\} = \varnothing$, then surely $- \langle x \rangle \cap \langle y \rangle = \varnothing$.

3.2 Sequention – Function

The basic idea (key point) of conversion of sequentions into functions is to concretize an abstract status of elements. All elements are considered to be Boolean variables. As a result sequention $\langle x_1 \, x_2 \dots x_n \rangle$ may be used as a format for generating the corresponding set of binary sequences. In the framework of this set, sequention is defined as a binary function with arguments, which are represented in the manner of sequences of binary elements-variables. There is the following dependence:

$$\langle x \rangle = F\left(\langle x_1\, x_2 \dots x_n \rangle\right). \tag{3.2}$$

All possible states of the function are $\langle x \rangle = 1$ and $\langle x \rangle = 0$. Sequention takes the value that depends on the values of variables, as well as, on the order how these values appear. Functional features of sequention are determined on the following grounds.

Sequention is able to take a unity value on condition that all its variables are also unity: $x_1 = 1$, $x_2 = 1$, and so on until $x_n = 1$. This statement can be expressed by means of conjunction operation in the compact logical form:

$$\bigwedge_{i=1}^{n} x_i = (x_1 \wedge x_2 \wedge \dots x_n) = 1. \tag{3.3}$$

The equation represents a necessary condition for sequention to be set at logic 1. However, it is not sufficient to have unity values of all variables for obtaining $\langle x \rangle = 1$. Corresponding binary set needs to be ordered in time taking into account priorities of elements in a sequence $\langle x_1\, x_2\, \dots\, x_n \rangle$. Concretely, logical unity setting $x_1 = 1$ needs to be ahead in time of all other settings: $x_2 = 1$, $x_3 = 1$, $x_4 = 1$, $\dots\, x_n = 1$. In turn, $x_2 = 1$ value needs to be ahead of $x_3 = 1$, $x_4 = 1$ $\dots x_n = 1$; $x_3 = 1$ – ahead of $x_4 = 1$ $\dots x_n = 1$, and so on.

The required sequencing of unity settings of sequention elements is displayed by binary relations of the following kind: $(x_1 = 1) \prec (x_2 = 1)$, $(x_2 = 1) \prec (x_3 = 1)$, $(x_3 = 1) \prec (x_4 = 1)$ and so on. Logical unity value of sequention may be obtained only in the case, when the rule of order relation $(x_i = 1) \prec (x_{i+1} = 1)$ is valid for all pairs of adjacent elements.

Binary sequences of logical unities and zeros ordered in accordance with a format $\langle x_1\, x_2\, \dots\, x_n \rangle$ represent the codomain of sequention. In this connection it is correct to consider the sequence $\langle 111 \dots 1 \rangle$ as a solution of the following equation:

$$F\left(\langle x_1\, x_2 \dots x_n \rangle\right) = 1. \tag{3.4}$$

Understanding of setting of one or another value of a binary variable is inextricably related with the switchings 0/1 and 1/0. Therefore, with reference to sequention it will be reasonable to agree with the solution presented in the following form:

$$\langle x \rangle = F\left(\langle (J/1)\,(0/1)\,(0/1) \dots (0/1) \rangle\right) = 1, \tag{3.5}$$

where $x_1 = J/1$, $x_2 = 0/1$, $x_3 = 0/1 \dots x_n = 0/1$.

From the obtained switching sequence follows that the setting $x_n = 0/1$ of the last variable immediately causes the unity setting of the sequention as a whole, that is $\langle x \rangle = 0/1$.

When the ordered setting of unity values is not performed, sequention stays in zero condition. Sequention uniquely takes a zero value, if at least one of variables is reduced to zero.

3.2.1 Definition of Sequention

Sequention is a function which takes 1 or 0 value depending on the values of binary arguments-variables as well as on the switching order of these variables. It is assumed that proposed binary operation of sequention is realized by a binary variable z as follows:

$$z = F(\langle x \rangle). \tag{3.6}$$

In view of the above-mentioned principles of sequention functioning, let us define that the corresponding logic obeys the following rules.

3.2.1.1 Unity Value Setting

- if $(x_i = 1) \prec (x_j = 1)$ for all i < j, then $z = 1$;
- if $(x_i = 1) \prec (x_{i+1} = 1)$ for all i < (n-1), then $z = 1$;
- if $\langle x_i\, x_{i+1} \rangle = 1$ for all i < (n-1), then $z = 1$;
- if $\langle x_1\, x_2\, \ldots\, x_n \rangle = \langle 11\ldots1 \rangle$, then $z = 1$.

3.2.1.2 Zero Value Setting

- if $(x_i = 1) \succ (x_j = 1)$ at least for one i < j, then $z = 0$;
- if $(x_i = 1) \succ (x_{i+1} = 1)$ at least for one i < (n-1), then $z = 0$;
- if $x_i = 0$ at least for one $i \in \{1, 2 \ldots n\}$, then $z = 0$;
- if $(x_1 \wedge x_2 \ldots x_n) = 0$, then $z = 0$.

3.2.1.3 Value Switchings

- if $\langle x_1\, x_2\, \ldots\, x_n \rangle = \langle (J/1)\, (0/1\, \ldots\, (0/1) \rangle$, then $z = 0/1$;
- if $\langle x_1\, x_2\, \ldots\, x_{n-1} \rangle = 1$ and $x_n = 0/1$, then $z = 0/1$;
- if $x_i = 1/0$ (any $i \in \{1, 2 \ldots n\}$ during $z = 1$, then $z = 1/0$.

Note

In consequence of the given definition the following rules are true:

- if $\langle x \rangle = 1$ and $\langle y \rangle \subseteq \langle x \rangle$, then $\langle y \rangle = 1$;
- if $\langle x \rangle = 0$ and $\langle x \rangle \cap \langle y \rangle \neq \varnothing$, then $\langle y \rangle = 0$.

3.2.2 Sequention-Function in Comparison with Sequention-Set

Sequention considered as an ordered set, and sequention which is assumed to be a function differ each from other. And this is not because of terminology niceties (details), but in essence. The same designations such as symbols x_i and x_j on the one hand are perceived as set members, and on the other hand they are binary variables – arguments of the sequential function. Therefore, it could not be excluded that nonempty set $\langle x \rangle \neq \varnothing$ generates functionally negligible sequention because of the permanent zero value $F(\langle x \rangle) = 0$. Exactly such a situation is provoked by the equality $x_i = \overline{x}_j$, a possibility of which is caused due to binary nature of the sequention elements.

The mentioned circumstance is revealed when comparing intersection and union operations of sets on the one hand and logical operations of conjunction and disjunction on the other hand. For example, the union $\langle x \rangle \cup \langle y \rangle$ of nonempty sets does not guarantee the effective conjunction; zero result $\langle x \rangle \wedge \langle y \rangle = 0$ is quite possible.

Another example:

-   if $\langle x \rangle$ is contained in $\langle y \rangle$ as a subset, equalities $\langle x \rangle \cap \langle y \rangle = \langle x \rangle$ and $\langle x \rangle \cup \langle y \rangle = \langle y \rangle$ are true.

In consequence of that these sequentions might be reduced to zero functionality, the given equalities not always ensure the result of conjunction and disjunction operations in the following forms: $\langle x \rangle \vee \langle y \rangle = \langle x \rangle$ and $\langle x \rangle \wedge \langle y \rangle = \langle y \rangle$. However, if the expected logical actions really take place, the corresponding positive results of intersections and unions are ensured.

3.2.3 Correct Sequentions

Taking into account performed above analysis, and intending to construct functionally valid sequentions it is necessary to introduce some restrictive measures. In this connection, all sequences, which contain elements related with equality $x_i = \overline{x}_j$ should be excluded from consideration. Any pair of elements (x_i, x_j) of the sequence $\langle x \rangle$ should obey the following inequalities: $\tilde{x}_i \neq \tilde{x}_j$, where $\tilde{x} \in \{x, \overline{x}\}$.

Thus, a principle of uniqueness of elements accepted above for sequention-set is also expanded to binary variables of sequention-function.

Sequentions without repeating elements (binary variables) are correct sequentions. Incorrect sequentions are functionally imperfect. Therefore, if it is not essential from the viewpoint of mathematics and analysis, let us hereinafter consider by default that all sequentions are correct and without inverse elements.

3.3 Simple and Complicated Sequentions

Sequentions by their structures may be simple and complicated, and in turn simple sequentions are subdivided into elementary and composite sequentions.

Elementary sequention $\langle x \rangle$ contains one sequence of the initial elements – binary variables. In some cases, it is allowed to represent single elements in the form of sequention as well, for example $\langle x_i \rangle$.

Two and more sequences form a simple composite sequention. For example, $\langle x\,y \rangle$ is a composite sequention, which consists of two sequences contained in elementary sequences $\langle x \rangle$ and $\langle y \rangle$.

Complicated sequentions differ in that they contain elements, which themselves are simple or complicated sequentions. For example, $\langle\langle x \rangle\,y \rangle$ is a complicated two-component sequention, which contains a simple sequention $\langle x \rangle$ followed by elementary sequence of the sequention $\langle y \rangle$. Expressions $\langle x\,\langle y \rangle\rangle$ and $\langle\langle x \rangle\,\langle y \rangle\rangle$ represent two-component sequentions as well.

Remark

Understanding of "element" in composite sequentions is interpreted specifically, that means – more widely. It is implied that "component" is considered as a certain complicated element. For example, from the sequention $\langle\langle x \rangle\langle y \rangle\rangle$ standpoint, its components $\langle x \rangle$ and $\langle y \rangle$ are elements with all the consequences that follow from this circumstance including the binary relation $\langle x \rangle \prec \langle y \rangle$. In accordance with the concept assumed, sequention $\langle x\,\langle y \rangle\rangle$ contains two equal in status elements; one of them is presented in the form of sequence $\langle y \rangle$, and another in the form of initial elements – binary variables $x_1, x_2 \dots x_n$.

3.3.1 Embedded Sequentions

Components, sequences of elements, elements themselves including elements-variables contained in the certain sequention; all of them are considered to be embedded in this sequention. For example, the following complicated sequention:

$$\langle\langle x\,\langle y\,z \rangle\rangle\,\langle\langle u \rangle\,\langle\langle v \rangle\,\langle p\,\langle r \rangle\rangle\,s \rangle\rangle\rangle \tag{3.7}$$

consists of eight simple elementary sequences: $\langle x \rangle$, $\langle y \rangle$, $\langle z \rangle$, $\langle u \rangle$, $\langle v \rangle$, $\langle p \rangle$, $\langle r \rangle$, $\langle s \rangle$. These sequences are embedded, but their embeddings are performed not equally. To be precise, $\langle x \rangle$, $\langle y \rangle$, $\langle z \rangle$, $\langle p \rangle$ and $\langle s \rangle$ are embedded in the form of elementary sequences, and $\langle u \rangle$, $\langle v \rangle$, $\langle r \rangle$ – in the form of component-sequentions. Sequences of initial elements of sequences $\langle y \rangle$ and $\langle z \rangle$ are parts of the simple composite sequention $\langle y\,z \rangle$; components $\langle p \rangle$ and $\langle r \rangle$ form a complicated two-component sequention $\langle p\,\langle r \rangle\rangle$. Sequention $\langle x\,\langle y\,z \rangle\rangle$ is formed by embedding sequence $\langle x \rangle$ and $\langle y\,z \rangle$ components. Sequentions $\langle v \rangle$ and $\langle p\,\langle r \rangle\rangle$, as well as sequence $\langle s \rangle$, are embedded into sequention $\langle\langle v \rangle\,\langle p\,\langle r \rangle\rangle\,s \rangle$, which in turn and together with $\langle u \rangle$ is a part of the next sequention $\langle\langle u \rangle\,\langle\langle v \rangle\,\langle p\,\langle r \rangle\rangle\,s \rangle\rangle$. The last sequention, as well as $\langle x\,\langle y\,z \rangle\rangle$, represents embeddings of the entire complicated sequention.

3.3.2 Embedding Layers of Sequentions

Embedding layers of sequentions are intended for ranking of components, which constitute one or another composite sequention. Offered for these purposes structure is built as a tree-type. Above sequention (3.7) includes 5 layers.

Layer 0

Zeroes layer is the layer assigned for function. It is occupied by entire sequention with all its embeddings.

Layer 1

On the first layer, the following components are embedded into initial sequention:

- complicated sequention $\langle x \langle y\, z \rangle \rangle$;
- complicated sequention $\langle \langle u \rangle \langle \langle v \rangle \langle p \langle r \rangle \rangle\, s \rangle \rangle$.

Layer 2

On the second layer, the following components are embedded into complicated sequentions of the first layer:

- initial elements $x_1, x_2 \ldots x_L$ of the sequention $\langle x \rangle$;
- simple composite sequention $\langle y\, z \rangle$;
- elementary sequention $\langle u \rangle$;
- complicated sequention $\langle \langle v \rangle \langle p \langle r \rangle \rangle\, s \rangle$.

Layer 3

On the third layer, the following components are embedded into sequentions of the second layer:

- initial elements $y_1, y_2 \ldots y_L$ of the sequention $\langle y \rangle$;
- initial elements $z_1, z_2 \ldots z_L$ of the sequention $\langle z \rangle$;
- initial elements $u_1, u_2 \ldots u_L$ of the sequention $\langle u \rangle$;
- elementary sequention $\langle v \rangle$;
- complicated sequention $\langle p \langle r \rangle \rangle$;
- initial elements $s_1, s_2 \ldots s_L$ of the sequention $\langle s \rangle$.

Layer 4

On the fourth layer, the following components are embedded into sequentions of the third layer:

- initial elements $v_1, v_2 \ldots v_L$ of the sequention $\langle v \rangle$;
- initial elements $p_1, p_2 \ldots p_L$ of the sequention $\langle p \rangle$;
- elementary sequention $\langle r \rangle$.

Layer 5

This layer is occupied by the following elements:

- initial elements $r_1, r_2 \ldots r_L$ of the sequention $\langle r \rangle$.

3.4 Binary Relations in Composite Sequentions

Order of variables arranged in composite sequentions is determined in the same manner as arrangement in elementary sequentions; that is, on the basis of binary relations between adjacent elements (components). Various kinds of these relations are presented bellow, depending on arrangement of all components in united sequention.

3.4.1 Relations of Two-Component Sequentions

3.4.1.1 Sequention of Two Sequences

Composite sequention $\langle x\ y \rangle$ contains elements of two sequences. Sequence taken from sequention $\langle x \rangle$ follows another sequence taken from sequention $\langle y \rangle$. Sequences are combined on the basis of a binary relation represented in the form: $x \prec y$. It is also possible to characterize the composite sequention by the following elementary binary relation: $x_L \prec x_F$. Both found relations generate the same sequention, and in this sense these binary relations are equivalent, that can be expressed as follows:

$$(x \prec y) \approx (x_L \prec y_F). \tag{3.8}$$

3.4.1.2 Sequence and Sequention

Complicated sequention $\langle x\ \langle y \rangle \rangle$ combines an elementary sequence of sequention $\langle x \rangle$ with sequention $\langle y \rangle$. Component association is performed in accordance with relationship $x \prec \langle y \rangle$. Along with this, binary relations $x_L \prec \langle y \rangle$ and $x_L \prec y_L$ are also available for arranging components in the given order. In other words, valid arrangement of variables is equally held by each of the presented binary relations. These relationships ensure required order, and therefore are equivalent:

$$(x \prec \langle y \rangle) \approx (x_L \prec \langle y \rangle) \approx (x_L \prec y_L). \tag{3.9}$$

3.4.1.3 Sequention and Sequence

In complicated sequention $\langle \langle x \rangle\ y \rangle$, sequence $\langle y \rangle$ is preceded by sequention $\langle x \rangle$. Components are arranged in order caused by the relation $\langle x \rangle \prec y$. Aside from this, it may be affirmed that composite sequention is composed on the basis of binary relations $\langle x \rangle \prec y_F$ or $x_L \prec y_F$. All mentioned relations are equivalent:

$$(\langle x \rangle \prec y) \approx (\langle x \rangle \prec y_F) \approx (x_L \prec y_F). \tag{3.10}$$

3.4.1.4 Combination of Two Sequentions

In complicated sequention $\langle \langle x \rangle\ \langle y \rangle \rangle$, an arrangement of components is predetermined by relation of two sequentions in the form $\langle x \rangle \prec \langle y \rangle$. Similarly ordered placement is also held by the binary relation $x_L \prec y_L$. The following equivalence takes place:

$$(\langle x \rangle \prec \langle y \rangle) \approx (x_L \prec y_L). \tag{3.11}$$

3.4.2 Relations of Sequentions in Connection with Order Relation

Obtained above binary relations are the rules, which need to be satisfied. Otherwise, combined sequentions form not properly operating function because of its imperfection. Furthermore, equivalent relationships allow coming to general conclusions in relation to functionality of composite sequentions.

Bellow it is shown how order relations of two-component sequentions influence upon logical relationships of sequentions themselves. The following principles are found for sequentions, whose connections are characterized by equivalent binary relations of their elements.

3.4.2.1 Principle 1

Sequention $\langle x\, y \rangle$ contains elements, which are ordered in accordance with a binary relation $x_L \prec y_F$. The same order is also set for another sequention $\langle\langle x \rangle\, y \rangle$. It means that embedded components of sequentions obey relations, which are equivalent as follows:

$$(x \prec y) \approx (\langle x \rangle \prec y). \tag{3.12}$$

Due to the equivalence of orders functional possibilities of sequentions $\langle x\, y \rangle$ and $\langle\langle x \rangle\, y \rangle$ are identical:

$$\langle x\, y \rangle = \langle\langle x \rangle\, y \rangle. \tag{3.13}$$

3.4.2.2 Principle 2

Sequention $\langle x\, \langle y \rangle \rangle$ contains elements, which are ordered in accordance with a binary relation $x_L \prec y_L$. The same order is also set for another sequention $\langle\langle x \rangle\, \langle y \rangle \rangle$. Therefore, binary relations between components of both sequentions are equivalent:

$$(x \prec \langle y \rangle) \approx (\langle x \rangle \prec \langle y \rangle), \tag{3.14}$$

and compared sequentions are equal:

$$\langle x\, \langle y \rangle \rangle = \langle\langle x \rangle\, \langle y \rangle \rangle. \tag{3.15}$$

3.4.2.3 Principle 3

Transitivity principal presented in formulae intended for binary relations causes the following assertion:

- if $x_L \prec y_F$, then $x_L \prec y_L$ because $y_F \prec y_L$.

In other words, binary relation $x_L \prec y_F$ covers other relation $x_L \prec y_L$. Further, taking into account equivalent connections earlier obtained for two-component sequentions, the following assertion is inevitable:

–   if $x \prec y$ or $\langle x \rangle \prec y$, then $x \prec \langle y \rangle$ and $\langle x \rangle \prec \langle y \rangle$.

In accordance with the given assertions, binary relations are connected by a certain logical manner. This connection is adequately reflected in the relationships between sequentions so that these relations take the form of implications:

$$\langle x\,y \rangle \Rightarrow \langle x \langle y \rangle \rangle , \qquad\qquad (3.16.1)$$

$$\langle x\,y \rangle \Rightarrow \langle \langle x \rangle \langle y \rangle \rangle , \qquad\qquad (3.16.2)$$

$$\langle \langle x \rangle\,y \rangle \Rightarrow \langle x \langle y \rangle \rangle , \qquad\qquad (3.16.3)$$

$$\langle \langle x \rangle\,y \rangle \Rightarrow \langle \langle x \rangle \langle y \rangle \rangle . \qquad\qquad (3.16.4)$$

Consequence

From (3.16) it follows:

–   if $\langle x\,y \rangle = 1$ or $\langle \langle x \rangle\,y \rangle = 1$, then $\langle x\,\langle y \rangle \rangle = 1$ and $\langle \langle x \rangle\,\langle y \rangle \rangle = 1$;
–   if $\langle x\,\langle y \rangle \rangle = 0$ *or* $\langle \langle x \rangle\,\langle y \rangle \rangle = 0$, then $\langle x\,y \rangle = 0$ and $\langle \langle x \rangle\,y \rangle = 0$.

Note

Two-component sequentions could be considered as fragments (adjacent components) of more complicated sequentions. From order relations of adjacent components, corresponding relationships for entire sequention are formed. Therefore, above-obtained results are available for any sequention regardless of its complexity. Coincident elements-variables and similar orders are sufficient conditions for equality of compared sequentions.

3.4.3 *Features of Sequentions with Common Elements*

Some elements of one sequention can be the same as elements contained in another sequention. For example, if $\langle x \rangle \cap \langle y \rangle = \langle z \rangle$, then $\langle z \rangle$ is a common sequention because $\langle z \rangle \subset \langle x \rangle$ and $\langle z \rangle \subset \langle y \rangle$. Thus, sequentions $\langle x \rangle$ and $\langle y \rangle$ are found to be functionally linked by sharing of common arguments.

For example, sequention $\langle \langle x \rangle\,\langle x\,y \rangle \rangle$ contains components $\langle x \rangle$ and $\langle x\,y \rangle$, which are mutually linked because of inclusion $\langle x \rangle \subset \langle x\,y \rangle$. Arrangement order is determined by the relation $\langle x \rangle \prec \langle x\,y \rangle$, which is reduced to the equivalent binary relation $x_L \prec y_F$. But the same relation causes ranked location of elements for another sequention: $\langle x\,y \rangle$. Therefore, the following equality is valid:

$$\langle \langle x \rangle\,\langle x\,y \rangle \rangle = \langle x\,y \rangle . \qquad\qquad (3.17)$$

As a result, elementary sequention $\langle x \rangle$ "disappears"; it is absorbed by the next sequention $\langle x\, y \rangle$.

Another example: complicated sequention $\langle\langle x\, y\, z\, u \rangle\langle y\, u\, v \rangle\rangle$ contains two simple sequentions, which share common sequences embedded into $\langle y \rangle$ and $\langle u \rangle$. Elements of sequentions $\langle x\, y\, z\, u \rangle$ and $\langle y\, u\, v \rangle$ obey the relationships $x \prec y$, $y \prec z$, $z \prec u$ and $y \prec u$, $u \prec v$ respectively. In view of the equivalence:

$$(u \prec v) \approx (\langle x\, y\, z\, u \rangle \prec \langle y\, u\, v \rangle)\,, \tag{3.18}$$

existing binary relation $u \prec v$ predetermines the ranking of components embedded directly into the given complicated sequention. Hence, this sequention is fully characterized by the following set of binary relations: $x \prec y$, $y \prec z$, $z \prec u$, $u \prec v$. The same relations cause arrangement of elements in sequention $\langle x\, y\, z\, u\, v \rangle$ that attests to the following equality:

$$\langle\langle x\, y\, z\, u \rangle\langle y\, u\, v \rangle\rangle = \langle x\, y\, z\, u\, v \rangle\,. \tag{3.19}$$

3.5 Functionally Imperfect Sequentions

Introduced earlier (see Sect. 3.2.3) concept of functional perfection (imperfection) is directly related to that sequention is binary-valued function. Thanks to unity value, function is able to identify a certain series of variable switchings, and respond by zero value to all other switchings. When unity value is not ensured quite at all, sequention operating becomes meaningless, and sequention itself proves to be functionally imperfect.

3.5.1 Definition of Imperfection

Sequention $\langle x \rangle$ considered to be functionally perfect only in the event that equation $F(\langle x \rangle) = 1$ possesses solution. But if the possible values of function are restricted by a set $\{0, J\}$, sequention is imperfect. Imperfection is expressed in that none of the binary sequences is able to provide unity value for sequential function. Sequention is permanently zeroed or its value could not be determined in principle.

Generally the range of imperfect function is defined as $\{1, 0, J\}\backslash 1$ and can be marked by $\backslash 1$ that means "without 1". If both values 0 and 1 are not perfectly achieved, the function is reduced to J.

3.5.2 Examples of Imperfect Sequentions

A simplest case of imperfection is provoked by the component repetition, for example:

$$\langle\langle x \rangle\langle x \rangle\rangle = \backslash 1\,. \tag{3.20}$$

Indefiniteness of sequention $\langle\langle x\rangle\,\langle x\rangle\rangle$ is caused by equality $\langle x\rangle = \langle x\rangle$. This circumstance disagrees with binary relation $x \prec x$ considered as necessary condition for unity valued function.

Imperfection takes place when making attempts to combine in one sequention components $\langle x\,y\rangle$ and $\langle y\rangle$, where $\langle y\rangle \supset \langle x\,y\rangle$. In this case the corresponding functions could not be determined unambiguously, that is:

$$\langle\langle x\,y\rangle\,\langle y\rangle\rangle = \backslash 1 , \qquad\qquad (3.21.1)$$

$$\langle\langle y\rangle\,\langle x\,y\rangle\rangle = \backslash 1 . \qquad\qquad (3.21.2)$$

Here, indeterminacy of sequentions is caused by that functions are not able to become unity because of irresistible obstacle expressed in the form of incorrect and therefore not realizable binary relation $y \prec y$.

Some sequentions are found to be imperfect because of another reason. They are absolutely unusable for receiving unity values, and constantly stay at zero. For example:

$$\langle\langle x\,y\rangle\,\langle y\,x\rangle\rangle = 0 . \qquad\qquad (3.22)$$

Zero result is explained because a binary relation $x \prec y$ observed in the component $\langle x\,y\rangle$ is not compatible with opposite relation $y \prec x$, which is characteristic of the mirror sequention $\langle y\,x\rangle$. Therefore unity value $\langle x\,y\rangle = 1$ inevitably results in zero sequention $\langle y\,x\rangle = 0$, and vice versa. Generally speaking, all mirror embeddings are inconsistent with perfect sequentions.

Another, not so visibly expressed reason of sequention zeroing presents itself in the following example:

$$\langle\langle y\,u\,z\,v\rangle\,\langle x\,y\,z\rangle\rangle = 0 . \qquad\qquad (3.23)$$

Zero value is effect of discrepancy between binary relation $z \prec v$ characteristic of the component $\langle y\,u\,z\,v\rangle$ and relation $z \succ v$, which takes place because sequention $\langle x\,y\,z\rangle$ lags behind sequention $\langle y\,u\,z\,v\rangle$. If components within sequention are rearranged, mentioned zero imperfection disappears.

Note

It is obvious that sequention is functionally imperfect, if at least one of embedded sequentions is functionally imperfect as well.

3.5.3 Criterions for Functional Imperfection

Functional imperfection of sequention is identified with help of the following criterions.

3.5.3.1 Inverse Elements

Element-variable x_i of sequention $\langle \ldots x_i \ldots\rangle$ presents in other sequention $\langle \ldots \overline{x}_i \ldots\rangle$, but in inverse form.

Explanation

Any binary sequence which causes $\langle ... x_i ... \rangle = 1$ inevitably reduces sequention $\langle ... \overline{x}_i ... \rangle = 0$ to zero, and vice versa. Such logic is characteristic of zero conjunction:

$$\langle ... x_i ... \rangle \wedge \langle ... \overline{x}_i ... \rangle = 0 . \tag{3.24}$$

Therefore any sequention with both above components is doomed to functional imperfection as follows:

$$\langle \langle ... x_i ... \rangle \langle ... \overline{x}_i ... \rangle \rangle = 0 . \tag{3.25}$$

Composite construction of functionally perfect sequentions does not allow combining components with inverse elements.

3.5.3.2 Inconsistent Binary Relations of Elements

Pair of elements (x, y) of sequention $\langle ... x ... y ... \rangle$ is contained in other sequention $\langle ... y ... x ... \rangle$ in reverse order.

Explanation

Binary relation $x \prec y$ characteristic of one sequention is inconsistent with relation $y \prec x$ of another sequention. This inconstancy inevitably causes the following zero result:

$$\langle \langle ... x ... y ... \rangle \langle ... y ... x ... \rangle \rangle = 0 . \tag{3.26}$$

Combining components with opposite order relations of their common elements, functionally imperfect sequention is formed.

3.5.3.3 Inconsistent Relations of Components

Intermediate element x of sequention $\langle ... x ... y \rangle$ coincides with the last element of sequention $\langle ... x \rangle$.

Explanation

Combining these sequentions in such succession that $\langle ... x \rangle$ follows $\langle ... x ... y \rangle$, functionally imperfect sequention is formed:

$$\langle ... x ... y \rangle \langle ... x \rangle = 0 . \tag{3.27}$$

Zero value is explained due to the fact that binary relation $x \prec y$ of sequention $\langle ... x ... y \rangle$ is inconsistent with relation $y \prec x$, which regularizes components in composite sequention.

Remark

In the case of sequention $\langle\langle\ldots x\rangle\langle\ldots x\ldots y\rangle\rangle$ with reverse arrangement of components, symptoms of functional imperfection are not observed.

3.5.3.4 Unacceptable Binary Relations

The last element of sequention $\langle\ldots x\ldots y\rangle$ coincides with the last element of other sequention $\langle\ldots u\ldots y\rangle$.

Explanation

Combining mentioned sequentions, two dispositions and therefore different binary relations might be used. They are $\langle\ldots x\ldots y\rangle \prec \langle\ldots u\ldots y\rangle$ and $\langle\ldots x\ldots y\rangle \prec \langle\ldots u\ldots y\rangle$. None of the relationships is acceptable because of the identity $y \equiv y$. Combined sequentions are inconsistent. Since zeroing criterions are not detected composite sequention "swoons" in its indeterminate condition as follows:

$$\langle\langle\ldots x\ldots y\rangle\langle\ldots u\ldots y\rangle\rangle = \backslash 1. \tag{3.28}$$

3.5.4 Compatible Sequentions

Let us assume that functional imperfection of sequention is caused by incompatibility of its components. In this connection it should be considered that sequentions as embedded components are incompatible, if there is no way available for forming functionally perfect sequention by associating the corresponding elements. But if sequentions are compatible, they do not provoke functional imperfection phenomenon and can be combined without problem. The possibility exists of obtaining sequention with partly compatible components, when functionality of composite sequention depends on sequencing of combined components. It is obvious that sequentions without common elements-variables are compatible by definition.

3.6 Splitting and Splicing of Sequentions

3.6.1 Splitting

Any simple elementary sequention can be transformed into functionally equivalent complicated sequention. To this effect, the following actions should be performed.

Stage 1

Divide the initial sequention $\langle x_1\, x_2\ldots x_i\, x_{i+1}\ldots x_{n-1}\, x_n\rangle$ into two subsequentions and $\langle x_{i+1}\ldots x_{n-1}\, x_n\rangle$, where adjacent elements x_i and x_{i+1} are separated.

Stage 2

Add element x_i to sequention $\langle x_{i+1} \dots x_{n-1}\, x_n \rangle$, and doing so form sequention $\langle x_i\, x_{i+1} \dots x_{n-1}\, x_n \rangle$.

Stage 3

Merge formed above sequentions as follows:

$$\langle x_1\, x_2 \dots x_i \rangle \langle x_i\, x_{i+1} \dots x_{n-1}\, x_n \rangle. \tag{3.29}$$

Resulting sequention is equal to initial one because both sequentions are characterized by the identical binary relations and the common set of elements as well.

Enumerated actions in essence fix the rule of sequention splitting. In this connection it should be noted that on the second stage not only single element can be used as additive. For example, sequentions $\langle x_{i-1}\, x_i \rangle$ and $\langle x_2\, x_3 \dots x_i \rangle$ are also available. Set of allowed supplements is restricted by sequentions, whose last elements are equal to x_i. Otherwise, merged sequention loses binary relation $x_i \prec x_{i+1}$, and therefore its function could not coincide with the initial sequention.

It is significant that splitting rule is applicable to simple composite sequentions. For this case two transformations are equally available. They are:

$$\langle x\, y \rangle = \langle \langle x \rangle \langle x_L\, y \rangle \rangle, \tag{3.30.1}$$

$$\langle x\, y \rangle = \langle \langle x\, y_F \rangle \langle y \rangle \rangle. \tag{3.30.2}$$

3.6.2 Splicing

If it is possible to split sequention, naturally it can be spliced. Splicing rule is backward directed procedure in contrast with splitting. As a result simple composite sequention is obtained by joining two elementary sequentions with common elements.

For example, if sequentions $\langle x \rangle$ and $\langle y \rangle$ are compatible, and they share common element because $x_L = y_F$ ($x_n = y_1$), these sequentions are spliced as follows:

$$\langle \langle x_1\, x_2 \dots x_n \rangle \langle y_1\, y_2 \dots y_m \rangle \rangle = \langle x_1\, x_2 \dots x_n\, y_2 \dots y_m \rangle, \tag{3.31.1}$$

$$\langle \langle x_1\, x_2 \dots x_n \rangle \langle y_1\, y_2 \dots y_m \rangle \rangle = \langle x_1\, x_2 \dots x_{n-1}\, y_1\, y_2 \dots y_m \rangle. \tag{3.31.2}$$

Splicing procedure could be performed on the basis of common sequentions as well. For example, in the case of common $\langle y \rangle$ splicing is represented by the following transformation:

$$\langle \langle x\, y \rangle \langle y\, z \rangle \rangle = \langle x\, y\, z \rangle. \tag{3.32}$$

This equality is caused by binary relations $\langle x \rangle \prec \langle y \rangle$ and $\langle y \rangle \prec \langle z \rangle$. As a result of splicing repeated y-component disappears.

3.7 Sequential Laws

Sequention, as being logic-dynamical function of binary variables, possesses specific features, due to which this function differs from the related (similar) logical formations of Boolean algebra: conjunction, disjunction, and so on. The corresponding characteristics, rules and postulates are represented below.

3.7.1 Commutativity, Associativity and Distributivity

As well as to venjunction, commutativity is not inherent to sequention. For example:

$$\langle x\,y\,z \rangle \neq \langle y\,x\,z \rangle . \tag{3.33.1}$$

$$\langle\langle x\,y \rangle \langle z\,u \rangle\rangle \neq \langle\langle z\,u \rangle \langle x\,y \rangle\rangle . \tag{3.33.2}$$

Mainly, associativity is not inherent to sequention:

$$\langle x \langle y\,z \rangle\rangle \neq \langle\langle x\,y \rangle z \rangle , \tag{3.34.1}$$

$$\langle x\,y \rangle \neq \langle\langle x \rangle \langle y \rangle\rangle , \tag{3.34.2}$$

$$\langle x\,y \rangle \neq \langle x \langle y \rangle\rangle , \tag{3.34.3}$$

$$\langle\langle x \rangle \langle y \rangle\rangle \neq \langle\langle x \rangle y \rangle , \tag{3.34.4}$$

$$\langle x \langle y \rangle\rangle \neq \langle\langle x \rangle y \rangle . \tag{3.34.5}$$

However there are sequentions, which obey the law of associativity:

$$\langle x\,y \rangle = \langle\langle x \rangle y \rangle , \tag{3.35.1}$$

$$\langle\langle x \rangle \langle y \rangle\rangle = \langle x \langle y \rangle\rangle . \tag{3.35.2}$$

Distributivity is partly inherent to sequention. For example:

$$\langle\langle x \langle y \rangle\rangle \langle x \langle z \rangle\rangle\rangle = \langle x \langle y \rangle \langle z \rangle\rangle , \tag{3.36.1}$$

$$\langle\langle x\,y \rangle \langle y\,z \rangle\rangle = \langle x\,y\,z \rangle , \tag{3.36.2}$$

$$\langle\langle x \langle y \rangle\rangle \langle\langle y \rangle z \rangle\rangle = \langle x \langle y \rangle z \rangle , \tag{3.36.3}$$

$$\langle\langle x \langle y \rangle\rangle \langle\langle y \rangle \langle z \rangle\rangle\rangle = \langle x \langle y \rangle \langle z \rangle\rangle . \tag{3.36.4}$$

In other cases sequentions does not comply with distributivity because of inequalities:

$$\langle\langle x\,y \rangle \langle x\,z \rangle\rangle \neq \langle x \langle y \rangle \langle z \rangle\rangle , \tag{3.37.1}$$

$$\langle\langle x\,y \rangle \langle y\,z \rangle\rangle \neq \langle\langle x \rangle y \langle z \rangle\rangle , \tag{3.37.2}$$

$$\langle\langle x\,y \rangle \langle z\,y \rangle\rangle \neq \langle\langle x \rangle \langle z \rangle y \rangle . \tag{3.37.3}$$

The last inequality beside of uncoordinated binary relations is notable due to sequention $\langle\langle x\ y\rangle\langle z\ u\rangle\rangle$, which is imperfect because of its indeterminacy instead of 1:

$$\langle\langle x\ y\rangle\langle z\ y\rangle\rangle = \backslash 1 . \tag{3.38}$$

3.7.2 Zeroing Rule

If sequention $\langle x\rangle$ is contained in sequention $\langle y\rangle$, that means $\langle x\rangle \subset \langle\ y\rangle$, and inequality $x_L \neq y_L$ takes place, association of these sequentions on the basis of binary relation $\langle y\rangle \prec \langle x\rangle$ forms a complicated sequention, functionality of which is restricted because of its zero value. For example, assuming $\langle y\rangle = \langle x\ z\rangle$, the following equations are obtained:

$$\langle\langle x\ z\rangle\langle x\rangle\rangle = 0 , \tag{3.39.1}$$

$$\langle\langle x\ z\rangle x\rangle = 0 , \tag{3.39.2}$$

$$\langle x\ z\langle x\rangle\rangle = 0 . \tag{3.39.3}$$

Unity value proves to be unattainable because of the insuperable advancing $(x_L = 1) \prec (y_L = 1)$.

3.7.3 Absorption Rule

If sequention $\langle x\rangle$ is contained in sequention $\langle y\rangle$, that means $\langle x\rangle \subset \langle\ y\rangle$, and inequality $x_L \neq y_L$ takes place, association of these sequentions on the basis of binary relation $\langle x\rangle \prec \langle y\rangle$ forms a complicated sequention, which is reduced to $\langle y\rangle$. Thus, sequention $\langle x\rangle$ is found to be absorbed.

For example, assuming $\langle y\rangle = \langle x\ z\rangle$, the following formulac take place:

$$\langle\langle x\rangle\langle x\ z\rangle\rangle = \langle x\ z\rangle , \tag{3.40.1}$$

$$\langle x\langle x\ z\rangle\rangle = \langle x\ z\rangle . \tag{3.40.2}$$

Remark

Absorption rule is valid only if sequentions are functionally perfect. Otherwise expected absorption does not happen. For example:

$$\langle\langle x\rangle\ x\ z\rangle \neq \langle x\ z\rangle , \tag{3.41}$$

If assumed binary relation $\langle x\rangle \prec \langle y\rangle$ is changed for $\langle y\rangle \prec \langle x\rangle$, imperfect sequentions are formed:

$$\langle z\ x\langle x\rangle\rangle = \langle\langle z\ x\rangle\langle x\rangle\rangle = \backslash 1 . \tag{3.42}$$

3.7.4 Splicing Rule

If sequentions $\langle x \rangle$ and $\langle y \rangle = \langle u\, z \rangle$ are such that elements of sequence $\langle y \rangle$ are partly contained in $\langle x \rangle$ owing to $\langle u \rangle \subset \langle x \rangle$, and equality $u_L = x_L$ takes place, then the formed complicated sequention $\langle\langle x \rangle \langle y \rangle\rangle$ is reduced to a simple sequention by splicing sequences $\langle x \rangle$ and $\langle y \rangle$. Corresponding transformations are represented as follows:

$$\langle\langle x \rangle \langle y \rangle\rangle = \langle\langle x \rangle \langle u\, z \rangle\rangle = \langle x\, z \rangle \, . \tag{3.43}$$

3.7.5 Splitting Rule

Splitting is the reverse action of splicing. Therefore, if sequention $\langle x \rangle$ is conditionally presented in the form of three sequences $\langle y\, u\, z \rangle$, and if subset of elements is chosen such that $\langle v \rangle \supseteq \langle u \rangle$, where $v_L = u_L$, then initially simple sequention $\langle x \rangle$ can be transformed into a complicated sequention. The result of splitting procedure is presented in the following form:

$$\langle x \rangle = \langle y\, u\, z \rangle = \langle\langle y\, u \rangle \langle v\, z \rangle\rangle \, . \tag{3.44}$$

Postulates

1. Functionality of any sequention independent of its complexity is fully defined by the binary relations of adjacent components presented in the form of relations between initial elements of x_F and x_L type.
2. Logical unity setting (0/1) of sequention is immediately preceded by the same setting of the last variable of the last component. Sequential function switches at the moment and in the course of $x_L = 0/1$ type switching.
3. Logical zero setting (1/0) of sequention is immediately preceded by the same setting of any variable. Sequential function switches at the moment and in the course of switching $x_i = 1/0$, $i \in \{1, 2 \ldots n\}$.

3.8 Methods for Decomposition of Sequentions

Splitting of sequentions is in essence a decomposition procedure; one sequention generates two subsequentions, which are combined into a form of two-component sequention. It is reasonable to presume that subsequentions in turn are decomposable as well. Further actions of this sort will inevitably lead to final sequention, which will contain components of minimum length that means – two-element sequentions, named minisequentions. It is the end of decomposition because minisequentions do not "disappear" due to the following transformation:

$$\langle x_1\, x_2 \rangle = \langle\langle x_1 \rangle \langle x_1\, x_2 \rangle\rangle \, . \tag{3.45}$$

Generally, different scenarios of decomposition can take place and therefore there are several ways of solving the corresponding problem. Result to be expected is caused by the set of split elements and depends on concrete elements, which are chosen for performing splitting actions.

3.8.1 Non-systemized Splitting

According to this method of decomposition, elements used for sequence splitting are chosen without preassigned regulation. This selection is arbitrary, that means – non-systemized. As an example, it is proposed to perform decomposition by splitting of seven-element sequention $\langle x_1\ x_2\ x_3\ x_4\ x_5\ x_6\ x_7 \rangle$. The corresponding procedure could be presented as follows.

Stage 1

Initial splitting is produced between elements x_4 and x_5. In other words (and it will be not in contradiction with principle of performed operation), element x_4 is subject to splitting, that is, split element. As a result two subsequentions are formed inside of the entire sequence

$$-\quad \langle\langle x_1\ x_2\ x_3\ x_4 \rangle \langle x_4\ x_5\ x_6\ x_7 \rangle\rangle .$$

Stage 2

Splitting of elements x_3 and x_5 is performed, in consequence of which new embeddings appear, and sequention becomes more complicated one;

$$-\quad \langle\langle\langle x_1\ x_2\ x_3 \rangle \langle x_3\ x_4 \rangle\rangle \langle\langle x_4\ x_5 \rangle \langle x_5\ x_6\ x_7 \rangle\rangle\rangle .$$

Stage 3

Final stage of splitting affects elements x_2 and x_6. As a result of non-systemized splitting, complicated sequention is formed, three embedding layers of which are occupied by seven minisequentions within the following sequention:

$$-\quad \langle\langle\langle\langle x_1\ x_2 \rangle \langle x_2\ x_3 \rangle\rangle \langle x_3\ x_4 \rangle\rangle \langle\langle x_4\ x_5 \rangle \langle\langle x_5\ x_6 \rangle \langle x_6\ x_7 \rangle\rangle\rangle\rangle .$$

Structure of the resulting expression in essence reflects staged order of the splitting described above. Non-systemized splitting generates sequention in the form of complicated composition with irregular placed embeddings.

Note

During decomposition the first element of sequention is not subject to splitting; it is senseless because of the following equality:

$$\langle\langle x_L \rangle \langle x \rangle\rangle = \langle x \rangle . \tag{3.46}$$

3.8.2 Regular Splitting into Minisequentions

This decomposition is based on the same principles and transformations that are characteristic of the previous method. However, splitting procedure differs from the above in that it uses additional operation. Transformation based on the presented earlier (see Sect. 3.4.2, Eq. 3.15) equality $\langle\langle x\rangle\langle y\rangle\rangle = \langle x\langle y\rangle\rangle$ accomplishes every stage of splitting and doing so, new image of sequention is produced. Furthermore, regular splitting with separation of minisequentions is systemized procedure. Comparing with sequention order, all stages are inverse ranked. Elements are subject to splitting in the following sequence: $x_{n-1}, x_{n-2} \dots x_3, x_2$.

For further explanations let us return to the initially considered sequention

$$-\quad \langle x_1\, x_2\, x_3\, x_4\, x_5\, x_6\, x_7\rangle\,.$$

Decomposition procedure begins with splitting of element x_6. As a result a complicated two-component sequention is formed:

$$-\quad \langle\langle x_1\, x_2\, x_3\, x_4\, x_5\, x_6\rangle\langle x_6\, x_7\rangle\rangle\,.$$

This sequention by means of the additional operation takes the following form:

$$-\quad \langle x_1\, x_2\, x_3\, x_4\, x_5\, x_6\, \langle x_6\, x_7\rangle\rangle\,.$$

The obtained sequention differs from the initial one because it contains minisequention $\langle x_6\, x_7\rangle = \langle x_{n-1}\, x_n\rangle$ as separate embedding in contrast with other elements.

Further transformations are performed in stages by splitting of elements x_5, x_4, x_3, and x_2 with obligatory separating out minisequentions $\langle x_5\, x_6\rangle$, $\langle x_4\, x_5\rangle$, $\langle x_3\, x_4\rangle$, and $\langle x_2\, x_3\rangle$ respectively. As a result of all these actions, initial elementary sequention takes the required final form:

$$-\quad \langle\langle x_1\, x_2\rangle\langle x_2\, x_3\rangle\langle x_3\, x_4\rangle\langle x_4\, x_5\rangle\langle x_5\, x_6\rangle\langle x_6\, x_7\rangle\rangle\,.$$

The final expression contains $(n-1) = 6$ minisequentions, which are regularly and equally placed on the first layer of embeddings. Each minisequention includes adjacent elements of the initial sequention.

Note

According to the given method of decomposition the final sequentions are structured so that their representations can be generalized on the basis of the following formula:

$$\langle x\rangle = \langle\langle x_1\, x_2\rangle\langle x_2\, x_3\rangle\dots\langle x_{i-1}\, x_i\rangle\langle x_i\, x_{i+1}\rangle\dots\langle x_{n-2}\, x_{n-1}\rangle\langle x_{n-1}\, x_n\rangle\rangle\,. \qquad (3.47)$$

Compact form of the given formula is expressed as elementary sequention $\langle y\rangle = \langle y_1\, y_2\, \dots\, y_i\, y_{i+1}\, \dots\, y_{m-1}\, y_m\rangle$, where $y_i = \langle x_i\, x_{i+1}\rangle$, $m = n-1$.

3.8.3 *Separation of Elements*

This method is based on the proved above (see Sect. 3.4.2, Eq. 3.13) equality $\langle x\, y\rangle = \langle\langle x\rangle\, y\rangle$, by means of which the initial sequention takes the following form:

$$\langle x_1\, x_2\, x_3\, \dots\, x_{n-1}\, x_n\rangle = \langle\langle x_1\, x_2\, x_3\, \dots\, x_{n-1}\rangle x_n\rangle\,. \qquad (3.48)$$

As it can be seen, due to the transformation the last element x_n becomes separated from all other elements of sequention. Thus, the first stage of decomposition is completed. Further transformations are connected with systemized separations of elements x_{n-1}, x_{n-2}, and so on until x_2.

After performing all announced actions a multi-layer and regular structured sequention is formed as follows: $\langle\langle\langle...\langle\langle\langle x_1 \rangle x_2 \rangle x_3 \rangle...\rangle x_{n-1} \rangle x_n \rangle$. So decomposition is accomplished.

3.9 Forms for Representation of Sequentions

Binary function presented by sequention can be adequately displayed in the form, which is available for its realization by means of usual operations of Boolean algebra and additional venjunctive operation as well.

Aside from sequention which is defined as ordered set of binary variables and represented in the corresponding form (see Sect. 3.1, Eq. 3.1), and sequention which is represented in the form of a sequence of minisequentions (see Sect. 3.8.2), an elementary sequention has conjunctive and venjunctive forms for its representation.

3.9.1 Relation of Sequention with Venjunction

In definition of sequention (see Sect. 3.2.1), a condition of advancing logical unity setting of one variable in relation to another variable is used. In accordance with this condition, the following rule is applied:

- Setting $x_j = 1$ is preceded by $x_i = 1$ if, and only if i < j.

Thereby sequention is associated with venjunction because it is obvious that mentioned above rule is in agreement with a relation $(x_i = 1) \prec (x_j = 1)$, which in essence designates binary switching $x_j = 0/1$ on the background $x_i = 1$. But "*switching... on the background*" is truly venjunction represented by the formula $x_j \angle x_i = 1$. Thus, venjunction is validly displayable in the form of two-element sequention that is minisequention:

$$(x_j \angle x_i) = \langle x_i \, x_j \rangle . \tag{3.49}$$

Two-component sequention holds connection with venjunction owing to the following equalities:

$$\langle\langle x \rangle \langle y \rangle\rangle = \langle y \rangle \angle \langle x \rangle , \tag{3.50.1}$$

$$\langle x \langle y \rangle\rangle = \langle y \rangle \angle \langle x \rangle . \tag{3.50.2}$$

As to sequentions $\langle x \, y \rangle$ and $\langle\langle x \rangle \, y \rangle$, similar equalities are also possible, but provided that sequence $\langle y \rangle$ consists of a single element. Then:

$$\langle x\, y_1 \rangle = y_1 \angle \langle x \rangle\,, \tag{3.51.1}$$

$$\langle \langle x \rangle\, y_1 \rangle = y_1 \angle \langle x \rangle\,. \tag{3.51.2}$$

In any other cases that sequention $\langle y \rangle$ contains at least two elements ($|y| \geq 2$), analogous transformations could not be obtained because of the following inequalities:

$$\langle x\, y \rangle \neq \langle y \rangle \angle \langle x \rangle\,, \tag{3.52.1}$$

$$\langle \langle x \rangle\, y \rangle \neq \langle y \rangle \angle \langle x \rangle\,. \tag{3.52.2}$$

3.9.2 Conjunctive Form

Judging by definition of sequention (see Sect. 3.2.1), and taking into account decomposition by splitting (see Sect. 3.8.2), sequential function is expressed via minisequentions, which are connected by logical operation – conjunction. Thus, sequention $\langle x \rangle$ is separated into its forming components as follows:

$$\langle x_1\, x_2\, x_3 \ldots x_{n-1}\, x_n \rangle = \langle x_1\, x_2 \rangle \wedge \langle x_2\, x_3 \rangle \wedge \ldots \langle x_{n-1}\, x_n \rangle\,. \tag{3.53}$$

Conjunctive form for representation of sequention is generalized by the following expression:

$$\bigwedge_i (x_i\, x_{i+1}),\ \ i \in \{1, 2 \ldots \text{n-1}\}\,. \tag{3.54}$$

Note

In the case of decomposition, obtained on the basis of non-systemized splitting of elements (see Sect. 3.8.1), conjunctive representation looks differently, for example:

$$-\quad (((\langle x_1\, x_2 \rangle \wedge \langle x_2\, x_3 \rangle) \wedge \langle x_3\, x_4 \rangle) \wedge ((\langle x_4\, x_5 \rangle \wedge ((\langle x_5\, x_6 \rangle \wedge \langle x_6\, x_7 \rangle)))\,.$$

Owing to the law of associativity, this conjunction is not regular, but it easy reduces to the conjunctive form of systematic kind (see Eq. 3.54) that is reasonably expected.

It is obvious that minisequention and venjunction are two analytical forms of functionally equivalent operations. Therefore, aside from conjunctive representations with the given above formulae the following representative varieties are available as well.

For example:

$$\bigwedge_i (x_{i+1} \angle x_i) = \tag{3.55}$$

$$= (((x_7 \angle x_6) \wedge (x_6 \angle x_5)) \wedge (x_5 \angle x_4)) \wedge ((x_4 \angle x_3) \wedge ((x_3 \angle x_2) \wedge (x_2 \angle x_1)))\,.$$

3.9.3 Venjunctive Form

According to developed above (see Sect. 3.8.3) method of decomposition, procedure of consequently performed separation of elements obeys the following rule:

$$\langle x_1\, x_2\, x_3 \ldots x_{n-1}\, x_n \rangle = \langle\langle\langle \ldots \langle\langle x_1\, x_2 \rangle x_3 \rangle \ldots \rangle x_{n-1} \rangle x_n \rangle . \tag{3.56}$$

Resulting sequention is represented in such a form, which allows introducing venjunctive operation in it. For this purpose it is sufficient to take into account above-mentioned logical relations between sequention and venjunction, and apply the following rule:

$$\langle\langle y \rangle x_i \rangle = x_i \angle \langle y \rangle . \tag{3.57}$$

In the context of the representative form this rule should be modified as follows:

$$\langle\langle x_1\, x_2 \ldots x_{i-1} \rangle x_i \rangle = x_i \angle \langle x_1\, x_2 \ldots x_{i-1} \rangle . \tag{3.58}$$

Thus, after required transformations initial sequention eventually takes the announced venjunctive form: $x_n \angle (x_{n-1} \angle (\ldots(x_3 \angle (x_2 \angle x_1))\ldots))$. The corresponding *cascade* function is represented in [2].

Findings

1. Functional logic of any simple sequention can be presented in systemized conjunctive form, in venjunctive form as well as in multi-version non-systemized conjunctive form.
2. Owing to above-cited formulae, sequention appears in more accustomed form of logical operations on binary variables.
3. Representative forms and conjunctive decomposition especially are such algebraic expressions, which are able to lay the justified foundations for involving sequentions into logical formulae as objects adapted to manipulations, which are characteristic of usual Boolean expressions.

3.10 Boolean Operations with Sequentions

As applying conjunction, the following rules of zeroing and idempotency are true:

$$\langle x \rangle \wedge \overline{\langle x \rangle} = 0 . \tag{3.59.1}$$

$$\langle x \rangle \wedge \langle x \rangle = \langle x \rangle , \tag{3.59.2}$$

Action of disjunction does not distinct from analogous operations with Boolean variables, and therefore obeys the following equalities:

$$\langle x \rangle \vee \langle x \rangle = \langle x \rangle, \tag{3.60.1}$$

$$\langle x \rangle \vee \overline{\langle x \rangle} = 1. \tag{3.60.2}$$

Logical negation of sequention is performed in two steps. At first, sequention should be represented in conjunction form (see Sect. 3.9.2) with minisequentions. Then keeping in view functional equivalence between minisequention and venjunction (see Sect. 3.9.3), it is required to transform found minisequentions using the following equality:

$$\overline{\langle x_i \, x_j \rangle} = \langle \overline{x_i} \rangle \vee \langle \overline{y_j} \rangle \vee \langle x_j \, x_i \rangle, \tag{3.61}$$

which in essence copies the corresponding formula assigned earlier (see Sect. 1.10.2, Eq. 1.34.1) for inverse venjunction.

Taking into account all required operations, logically negated sequention is found as a result of the following transformations:

$$\overline{\langle x_1 \, x_2 \, x_3 \ldots x_{n-1} \, x_n \rangle} = \overline{\langle x_1 \, x_2 \rangle} \vee \overline{\langle x_2 \, x_3 \rangle} \vee \ldots \overline{\langle x_{n-1} \, x_n \rangle} = \tag{3.62}$$

$$= \overline{x_1} \vee \overline{x_2} \vee \overline{x_3} \vee \ldots \overline{x}_{n-1} \vee \overline{x}_n \vee \langle x_2 \, x_1 \rangle \vee \langle x_3 \, x_2 \rangle \vee \ldots \langle x_n \, x_{n-1} \rangle.$$

Inverse sequention obeys the rule expressed in the compact form as follows:

$$\overline{\langle x \rangle} = \bigvee_i \overline{x}_i \bigvee_j \langle x_j \, x_{j-1} \rangle, \tag{3.63}$$

where $i \in \{1, 2 \ldots n\}, j \in \{2, 3 \ldots n\}$.

Minimum length sequention (minisequention) in essence is a functional equivalent of venjunction. Therefore in the case of mirror minisequentions, it is reasonable to consider that result of disjunction coincides with earlier (see Sect. 1.10.3, Eq. 1.36.1) found, that is:

$$\langle x_1 \, x_2 \rangle \vee \langle x_2 \, x_1 \rangle = x_1 \wedge x_2. \tag{3.64}$$

As to conjunction, this operation on mirror sequentions always results in zero.

While performing conjunctive and disjunctive operations, sequentions are able to absorb each other. As most significant examples the following equalities are presented:

$$\langle x \, y \rangle \wedge \langle \langle x \rangle \langle y \rangle \rangle = \langle x \, y \rangle, \tag{3.65.1}$$

$$\langle \langle x \rangle \, y \rangle \rangle \wedge \langle x \langle y \rangle \rangle = \langle \langle x \rangle \, y \rangle, \tag{3.65.2}$$

$$\langle x \, y \rangle \vee \langle \langle x \rangle \langle y \rangle \rangle = \langle \langle x \rangle \langle y \rangle \rangle, \tag{3.65.3}$$

$$\langle \langle x \rangle \, y \rangle \rangle \vee \langle x \langle y \rangle \rangle = \langle x \langle y \rangle \rangle. \tag{3.65.4}$$

3.10.1 General Principles for Conjunctions

Principle 1

If at least one of elements of sequention $\langle x \rangle$ is presented by binary variable in its direct (without negation) form, and the same variable is contained in another sequention $\langle y \rangle$ as inversion, the result of conjunction will be equal to logical zero.

Principle 2

If ranked order of variables in sequention $\langle x \rangle$ is in contradiction with a sequence $\langle y \rangle$, conjunction $\langle x \rangle \wedge \langle y \rangle$ reduces to zero. For inconsistency condition it is sufficient to have at least one pair of elements, which are inversely ordered in sequentions $\langle x \rangle$ and $\langle y \rangle$.

Principle 3

If one sequention is included into another one as subset, conjunction of these sequentions is performed with absorption, where included sequention is absorbed by including sequention. For example, in the case $\langle y \rangle \subset \langle x \rangle$, conjunction $\langle x \rangle \wedge \langle y \rangle$ becomes sequention $\langle x \rangle$. Conjunction provides sequention with ability to absorb any of its part.

Principle 4

If sequentions $\langle x \rangle$ and $\langle y \rangle$ are compatible and additionally contain common elements, that means $\langle x \rangle \cap \langle y \rangle \neq \varnothing$, then in the case of the common component $\langle z \rangle$ inclusion in the form $\langle z \rangle \subset \langle x \rangle$ and $\langle z \rangle \subset \langle y \rangle$, the following variants are possible:

- if $\langle x \rangle = \langle z\,u \rangle$ and $\langle y \rangle = \langle z\,v \rangle$, then $\langle z\,u \rangle \wedge \langle z\,v \rangle = \langle z \rangle \wedge \langle z_L\,u \rangle \wedge \langle z_L\,v \rangle$;
- if $\langle x \rangle = \langle u\,z \rangle$ and $\langle y \rangle = \langle v\,z \rangle$, then $\langle u\,z \rangle \wedge \langle v\,z \rangle = \langle z \rangle \wedge \langle u\,z_F \rangle \wedge \langle v\,z_F \rangle$;
- if $\langle x \rangle = \langle z\,u \rangle$ and $\langle y \rangle = \langle v\,z \rangle$, then $\langle z\,u \rangle \wedge \langle v\,z \rangle = \langle z \rangle \wedge \langle z_L\,u \rangle \wedge \langle v\,z_F \rangle$
- if $\langle x \rangle = \langle u\,z \rangle$ and $\langle y \rangle = \langle z\,v \rangle$, then $\langle u\,z \rangle \wedge \langle z\,v \rangle = \langle z \rangle \wedge \langle u\,z_F \rangle \wedge \langle z_L\,v \rangle$.

3.10.2 General Principles for Disjunctions

Principle 1

If one sequention is included into another one as subset, disjunction of these sequentions is performed with absorption, where included sequention absorbs an including sequention. For example, in the case $\langle y \rangle \subset \langle x \rangle$ disjunction $\langle x \rangle \vee \langle y \rangle$ becomes sequention $\langle y \rangle$. Disjunction provides sequention with ability to be absorbed by any of its part.

Principle 2

If sequentions $\langle x \rangle$ and $\langle y \rangle$ contain common elements, that means $\langle x \rangle \cap \langle y \rangle \neq \varnothing$, then in the case of the common component $\langle z \rangle$ inclusion in the form $\langle z \rangle \subset \langle x \rangle$ and $\langle z \rangle \subset \langle y \rangle$, the following variants are possible:

- if $\langle x \rangle = \langle z\,u \rangle$ and $\langle y \rangle = \langle z\,v \rangle$, then $\langle z\,u \rangle \vee \langle z\,v \rangle = \langle z \rangle \wedge (\langle z_L\,u \rangle \vee \langle z_L\,v \rangle)$;
- if $\langle x \rangle = \langle u\,z \rangle$ and $\langle y \rangle = \langle v\,z \rangle$, then $\langle u\,z \rangle \vee \langle v\,z \rangle = \langle z \rangle \wedge (\langle u\,z_F \rangle \vee \langle v\,z_F \rangle)$;
- if $\langle x \rangle = \langle z\,u \rangle$ and $\langle y \rangle = \langle v\,z \rangle$, then $\langle z\,u \rangle \vee \langle v\,z \rangle = \langle z \rangle \wedge (\langle z_L\,u \rangle \vee \langle v\,z_F \rangle)$;
- if $\langle x \rangle = \langle u\,z \rangle$ and $\langle y \rangle = \langle z\,v \rangle$, then $\langle u\,z \rangle \vee \langle z\,v \rangle = \langle z \rangle \wedge (\langle u\,z_F \rangle \vee \langle z_L\,v \rangle)$.

3.10.3 General Principles for Venjunctions

Principle 1

Venjunction $\langle x \rangle \angle \langle y \rangle$ by its functional possibilities does not differs from two-component sequence $\langle \langle y \rangle \langle x \rangle \rangle$. This circumstance is embodied in the following equality: $\langle x \rangle \angle \langle y \rangle = \langle \langle y \rangle \langle x \rangle \rangle$.

Principle 2

If sequention $\langle y \rangle$ is included into another sequention $\langle x \rangle$ in the form $\langle y \rangle \subset \langle x \rangle$, then in the case $\langle x \rangle = \langle y\,z \rangle$ the following variants are possible.

1. In consequence of operation $\langle x \rangle \angle \langle y \rangle$, including sequention absorbs included sequention as follows:
   - $\langle x \rangle \angle \langle y \rangle = \langle y\,z \rangle \angle \langle y \rangle = \langle \langle y \rangle \langle y\,z \rangle \rangle = \langle y\,z \rangle = \langle x \rangle$.

2. In the case of mirror operation $\langle y \rangle \angle \langle x \rangle$, venjunction is found to be functionally imperfect. Corresponding transformations result in zero value:
   - $\langle y \rangle \angle \langle x \rangle = \langle y \rangle \angle \langle y\,z \rangle = \langle \langle y\,z \rangle \langle y \rangle \rangle = 0$.

Principle 3

If sequentions $\langle x \rangle$ and $\langle y \rangle$ contain common elements in the form of sequence $\langle z \rangle$, then in relation to venjunction $\langle x \rangle \angle \langle y \rangle$, the following results are possible.

Zeroing

If $\langle x \rangle = \langle u\,z \rangle$ and $\langle y \rangle = \langle z\,v \rangle$, then $\langle u\,z \rangle \angle \langle z\,v \rangle = \langle \langle z\,v \rangle \langle u\,z \rangle \rangle = 0$.

Indeterminacy

If $\langle x \rangle = \langle u\,z \rangle$ and $\langle y \rangle = \langle v\,z \rangle$, then $\langle u\,z \rangle \angle \langle v\,z \rangle = \langle \langle v\,z \rangle \langle u\,z \rangle \rangle = \backslash 1$.

Simple composite sequention

If $\langle x \rangle = \langle z\,u \rangle$ and $\langle y \rangle = \langle v\,z \rangle$, then $\langle z\,u \rangle \angle \langle v\,z \rangle = \langle \langle v\,z \rangle \langle z\,u \rangle \rangle = \langle v\,z\,u \rangle$.

Complicated sequention

If $\langle x \rangle = \langle z\,u \rangle$ and $\langle y \rangle = \langle z\,v \rangle$, then $\langle z\,u \rangle \angle \langle z\,v \rangle = \langle \langle z\,v \rangle \langle z\,u \rangle \rangle$.

3.11 Transformation of Complicated Sequentions

Transformations of complicated sequentions differ from analogous actions with simple sequentions at least due to the difference in type of elements. Initial elements are contained in simple sequentions; complicated sequentions contain components, that is, not only elements but its sequentions and subsequentions as well. Additionally these components are placed on several embedding layers.

Problems connected with representation of elementary sequentions were solved earlier (see Sect. 3.9.2, 3.9.3). Therefore transformation problem lies in actions, which should be performed for expansion of complicated sequentions into simple forming components. The problem characteristic only of complicated sequentions is modification that means transformation of components, addition of new elements, and change of embedding layers and the structure of sequention.

3.11.1 Conjunctive Expansion

Conjunctive representation of complicated sequention is based on the conjunctive form (see Sect. 3.9.2) of elementary sequention as well as on the rules assigned for expansion of two-component sequentions. This is because a simple sequention transforms into conjunction of minisequentions, that is, two-element sequentions. But in the case of complicated sequention the same two-element, precisely two-component subsequentions are subject to further transformation. Corresponding actions are specified by the following rules:

$$\langle x\, y \rangle = \langle \langle x \rangle\, y \rangle \Leftrightarrow \langle x \rangle \wedge \langle y \rangle \wedge \langle x_L\ y_F \rangle , \tag{3.66.1}$$

$$\langle \langle x \rangle \langle y \rangle \rangle = \langle x \langle y \rangle \rangle \Leftrightarrow \langle x \rangle \wedge \langle y \rangle \wedge \langle x_L\ y_L \rangle . \tag{3.66.2}$$

These expressions represent logical equivalence of two-component sequentions, on the one hand, and formulae obtained by conjunctive expansion of these sequentions on the other hand.

3.11.1.1 Example of Expansion of Sequention

For example, the following complicated sequential function is chosen from Sect. 3.3.1:

$$F(\langle x \rangle, \langle y \rangle, \langle z \rangle, \langle u \rangle, \langle v \rangle, \langle p \rangle, \langle r \rangle, \langle s \rangle) = \langle \langle x \langle y\, z \rangle \rangle \langle \langle u \rangle \langle \langle v \rangle \langle p \langle r \rangle \rangle s \rangle \rangle \rangle . \tag{3.67}$$

Three layers of the sequention are occupied with components, which properly speaking are subjects to expansion. If planned converting actions distribute in accordance with embedding layers (see Sect. 3.3.2), required procedure appears in the following systemized form.

Layer 3

$$-\quad \langle p \langle r \rangle \rangle = \langle p \rangle \wedge \langle r \rangle \wedge \langle p_L\ r_L \rangle .$$

Layer 2

- $\langle y\,z \rangle = \langle y \rangle \wedge \langle z \rangle \wedge \langle y_L\,z_F \rangle$,
- $\langle \langle v \rangle \langle p \langle r \rangle \rangle \rangle = \langle v \rangle \wedge \langle p \rangle \wedge \langle r \rangle \wedge \langle p_L\,r_L \rangle \wedge \langle v_L\,r_L \rangle$.

Layer 1

- $\langle x \langle y\,z \rangle \rangle = \langle x \rangle \wedge \langle y \rangle \wedge \langle z \rangle \wedge \langle y_L\,z_F \rangle \wedge \langle x_L\,z_L \rangle$,
- $\langle \langle u \rangle \langle \langle v \rangle \langle p \langle r \rangle \rangle s \rangle \rangle = \langle u \rangle \wedge \langle v \rangle \wedge \langle p \rangle \wedge \langle r \rangle \wedge \langle s \rangle \wedge \langle p_L\,r_L \rangle \wedge \langle v_L\,r_L \rangle \wedge \langle u_L\,r_L\,s_F \rangle$

In accordance with the layered scheme, components of the third layer are decomposed in the first place. Then the result obtained is taken into account on the next (second) layer, and planned transformations are performed with reference to components of this layer. And so on until the complete expansion on the first layer. Finally (on a zero layer) initial sequence is found to be presented in the following form:

$$F \left(\langle x \rangle, ... \langle s \rangle \right) = \langle x \rangle \wedge \langle y \rangle \wedge \langle z \rangle \wedge \langle u \rangle \wedge \langle v \rangle \wedge \langle p \rangle \wedge \langle r \rangle \wedge \langle s \rangle \wedge \qquad (3.68)$$

$$\wedge \langle x_L\,z_L \rangle \wedge \langle y_L\,z_F \rangle \wedge \langle u_L\,r_L\,s_F \rangle \wedge \langle v_L\,r_L \rangle \wedge \langle p_L\,r_L \rangle \wedge \langle z_L\,s_F \rangle .$$

The given conjunctive expansion is represented by two types of forming components that is characteristic of this operation.

1. Conjunction of elementary sequentions is responsible for the logical unity setting of all components.
2. Sequentions, which contains first and last elements of embedded components, control the succession of the mentioned settings.

Definitively, that means after conjunctive decomposition of simple sequentions, a complicated sequention takes the form of conjunction on minisequentions (see Sect. 3.9.2).

Remark

Let us recall the auxiliary function marked λ, which was introduced earlier (see Sect. 1.6.4) for definition of venjunction. In the light of obtained results, this function appears to be useful as a part of the following expansion of minisequention: $\langle x\,y \rangle = x \wedge y \wedge \lambda$. Thus, function λ is a prototype of those forming components, which are responsible for order of switchings in complicated sequentions.

3.11.2 Venjunctive Separation

A given transformation does not copy the analogous representative form (see Sect. 3.9.3), because that venjunctive form is applicable to simple sequentions only. Venjunctive separation is based on the relations of two-component sequentions with venjunctions (see Sect. 3.9.1, Eq. 3.49, Eq. 3.50). For example, a complicated sequention

$$- \quad \langle\langle\langle x\rangle\langle\langle y\rangle\langle z\rangle\rangle\rangle\langle\langle\langle u\rangle\langle v\rangle\langle p\rangle\rangle\langle r\rangle\rangle\rangle,$$

taking into account its three embedding layers is subdivided by venjunctive operation as follows:

Layer 1

$$- \quad \langle\langle\langle u\rangle\langle v\rangle\langle p\rangle\rangle\langle r\rangle\rangle \angle \langle\langle x\rangle\langle\langle y\rangle\langle z\rangle\rangle\rangle.$$

Layer 2

$$- \quad (\langle r\rangle \angle \langle\langle u\rangle\langle v\rangle\langle p\rangle\rangle) \angle (\langle\langle y\rangle\langle z\rangle\rangle \angle \langle x\rangle).$$

Layer 3

$$- \quad (\langle r\rangle \angle (\langle p\rangle \angle (\langle v\rangle \angle \langle u\rangle)))\angle ((\langle z\rangle \angle \langle y\rangle)\angle \langle x\rangle).$$

3.11.3 Modification of Sequentions

Modification is a way to transform sequention in order to obtain other functionally equivalent form of this sequention. Modification purposes may be various:

- to adapt sequentions for the discovered above operations of conjunctive expansion and venjunctive separation;
- to optimize sequentions taking into account their further realization in a form of the corresponding logical circuits;
- to minimize long sequences whose implementation is difficult because of existing restrictions in type or set of logical elements.

In the context of modification of complicated sequentions, embedded simple and composite elementary sequentions are also subjects of transformations.

Manipulations needed for transformation of sequentions, are performed in accordance with certain rules. These rules are represented by the following formulae:

$$\langle x\langle y\rangle\rangle => \langle\langle x\rangle\langle y\rangle\rangle, \tag{3.69.1}$$

$$\langle x\, y\rangle = \langle\langle x\rangle\, y\rangle => \langle\langle x\, y_F\rangle\langle y\rangle\rangle, \tag{3.69.2}$$

$$\langle x\, y\rangle = \langle\langle x\rangle\, y\rangle => \langle\langle x\rangle\langle x_L\, y\rangle\rangle, \tag{3.69.3}$$

$$\langle x\, y\rangle = \langle\langle x\rangle\, y\rangle => \langle\langle x\rangle\, y_F\, \langle y\rangle\rangle, \tag{3.69.4}$$

$$\langle x\, y\rangle = \langle\langle x\rangle\, y\rangle => \langle\langle x\rangle\langle x_L\, y_F\rangle\langle y\rangle\rangle. \tag{3.69.5}$$

Here, equality symbol together with arrow constitutes a sign $=>$, which indicates direction of transformation. On the left of this sign, initial sequentions are placed; on the right, modified forms of these sequentions are represented.

As a result of modification complicated sequentions lose embeddings with sequences of the initial variables. In such a manner sequences are transformed into sequentions. In certain cases first and last elements of other sequentions are involved.

For example, two modified sequentions are represented below:

$$\langle x\,y\,z\,\langle u\rangle\rangle = \langle\langle x\,y_F\rangle\langle y\rangle\langle y_L\,z\rangle\langle u\rangle\rangle\,, \qquad (3.70.1)$$

$$\langle x\langle y\,z\rangle\langle u\rangle\,v\,\langle r\rangle\rangle = \langle\langle x\rangle\langle\langle y\rangle\langle y_L\,z_F\rangle\langle z\rangle\rangle\langle u\rangle\langle u_L\,v_F\rangle\langle v\rangle\langle r\rangle\rangle\,. \qquad (3.70.2)$$

3.12 Graphics of Sequentions; Memory Depth and Volume

3.12.1 Graph of Sequention

Graphic image in Fig. 3.1 represents a complicated sequention, conjunctive expansion of which is found above in Sect. 3.11.1.1 (Eq. 3.68).

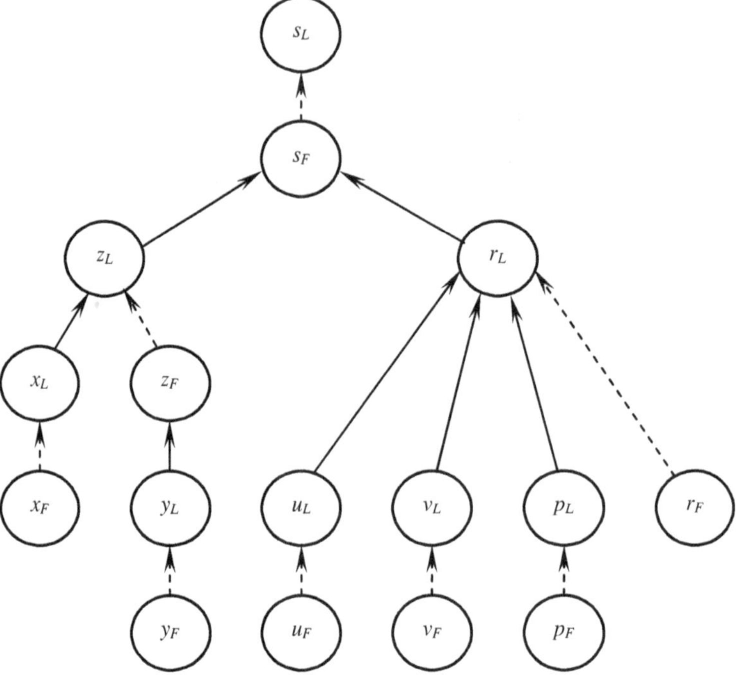

Fig. 3.1 Graph of example sequention $\langle\langle x\langle y\,z\rangle\rangle\langle\langle u\rangle\langle\langle v\rangle\langle p\,\langle r\rangle\rangle\,s\rangle\rangle\rangle$

Nodes of the graph are associated with elements of sequention. Arcs of the graph demonstrate how these elements are ordered. For example, connective $z_L \rightarrow s_L$ signifies that in accordance with a binary relation $z_L \prec s_L$ the last element of sequention $\langle z\rangle$ directly precedes the last element of sequention $\langle s\rangle$. Elementary sequentions are represented by their first and last elements, which are connected by dotted lines. Thereby not-shown intermediate elements are meant, for example, x_2, x_3 ... x_{L-1} between x_F and x_L in connective $x_F \; -- \rightarrow x_L$.

Note

There is one-to-one correspondence between sequention and its graph. All binary relations, which are characteristic of sequention, are displayed in graph. Graph does not represent binary relations, which are not caused by sequention.

3.12.2 Memory Depth

In the context of asynchronous logic, let us consider that memory depth of a binary sequential function is defined on the basis of the following thought.

Logical constants 0 and 1 do not depend on time. Constant is a permanent state, in reference to which any temporal categories including memory are improper. Taking into account this circumstance, absence of memory should be considered as zero memory. Logical constants possess zero memory that means zero depth: $Md = 0$.

Combinational functions of binary variables possess memory with a depth $Md = 1$. Corresponding realizations in the form of combinational circuits are able to "remember" their current state.

Memory of bistable cell and some triggers extends to the previous (passed) state of these devices. So memory depth $Md = 2$ is achieved. In the framework of asynchronous logic, such memory is characteristic of the simplest functions – minisequention and two-variable venjunction.

Functions with memory $Md = 3$ "remember" previous (last) and before last states. Functions with $Md = 4$ are able to memorize previous of "before last" state. And so on.

It is obvious that in asynchronous logic, numerical value of memory depth is associated with a length of sequention. For example, memory depth for elementary sequention $\langle x_1 \; x_2 \; ... \; x_n \rangle$ is equal to the length of this sequention, that is $Md = n$.

3.12.3 Memory of Complicated Sequentions

With reference to complicated sequentions, memory depth is calculated on the basis of certain rules, which are represented bellow by the example of the graph in Fig. 3.1.

The graph contains one output and six input nodes. Each input node via the corresponding arcs and intermediate nodes is connected with the output node. These connections, these paths are associated with sequences of elements and therefore could be represented in the form of the following simple sequentions:

- $\langle x_F \; ... \; x_L \, s_F \; ... \; s_L \rangle$,
- $\langle y_F \; ... \; y_L \, z_F \; ... \; z_L s_F \; ... \; s_L \rangle$,

- $\langle u_F \dots u_L\, r_L\, s_F \dots s_L \rangle,$
- $\langle v_F \dots v_L\, r_L\, s_F \dots s_L \rangle,$
- $\langle p_F \dots p_L\, r_L\, s_F \dots s_L \rangle,$
- $\langle r_F \dots r_L\, s_F \dots s_L \rangle.$

Memory depth of the given sequention, as well as sequention as such, is defined by the longest path in the graph. As to Fig. 3.1, memory depth is found in accordance with the following expression:

$$\mathrm{Md} = |s| + \max\left[(|x|+1),\, (|y|+|z|),\, (|u|+1),\, (|v|+1),\, (|p|+1),\, (|r|)\right]. \qquad (3.71)$$

3.12.4 Memory Volume

In sequential logic function of asynchronous memory is performed by an operation of venjunction. Minimal memory components are represented in a form of two-element venjunctions or in a form of minisequentions. From a functional standpoint these representative forms are equal (see Sect. 3.9.1, Eq. 3.49).

Memory volume (Mv) of sequention is assigned by a total number of the corresponding minimal components. For example, simple n-sequention has Mv = n-1.

In the case of complicated sequention, memory volume is calculated in view of the following order of all including embedded sequentions. So for complicated sequention, which is represented by the graph in Fig. 3.1, memory volume is calculated by the following formula:

$$\mathrm{Mv} = (|x|-1)+(|y|+|z|-1)+(|u|-1)+(|v|-1)+(|p|-1)+(|r|-1)+(|s|-1)+6.$$
$$(3.72)$$

Numeral 6 designates a number of memory elements, which are expressed in a form of supplemented minisequentions: $\langle x_L\, z_L \rangle$, $\langle z_L\, s_F \rangle$, $\langle u_L\, r_L \rangle$, $\langle v_L\, r_L \rangle$, $\langle p_L\, r_L \rangle$, and $\langle r_L\, s_F \rangle$. According to Sect. 3.12.3, exactly these minisequentions support necessary order relations, which are dictated by a structure of the initial sequention.

Note

After relay schemes and combinational circuits the theory of finite-state automata [3, 4] allows creating different models for sequential circuit representation and design [5, 6]. To realize necessary memory function, sequential operators are involved in the form of Muller C-gate [7], SR flip-flop, and other available logical elements. Developed and represented above logic introduces venjunction and sequention – asynchronous operational elements of new type.

References

[1] Vasyukevich, V.: Monotone sequences of binary data sets and their identification by means of venjunctive functions. ACCS Journal 32(5), 49–56 (1998)

[2] Vasyukevich, V.: Asynchronous sequences decoding. ACCS Journal 41(2), 93–99 (2007)

[3] Gill, A.: Introduction to the Theory of Finite-State Machines. McGraw-Hill, New York (1962)
[4] Yakubaitis, E., Vasyukevich, V., et al.: Automata theory. Journal of Mathematical Sciences 7(2), 193–243 (1977)
[5] McCluskey, E.: Introduction to the Theory of Switching Circuits. McGraw-Hill Book Company, New York (1965)
[6] Hill, F., Peterson, G.: Introduction to Switching Theory and Logical Design. John Wiley and Sons, Chichester (1968)
[7] Muller, D., Bartky, W.: A Theory of Asynchronous Circuits. In: Proc. Int'l Symp. Theory of Switching, Part 1, pp. 204–243 (1959)

Chapter 4
Circuit Design

Abstract. In the fourth chapter questions of analysis and synthesis of asynchronous sequential circuits are examined. In this connection a group of trigger-type devices is allocated. In general case they are represented by a certain venjunctive function of two variables. This function obeys the rules of operation of a bistable cell in its definitely restricted mode. To thoroughly characterize trigger devices a new concept is introduced in the form of memory formula, which is used alongside with setting operations. Also behavioral model of circuits with memory is developed. It is a zone model, basic configurations of which compose all possible transitions between setting zones and memory zone. Mathematical apparatus of venjunction is used for analysis of positive and negative feedbacks. Some ways are proposed for interruption of spurious oscillations and for retention of settings which are caused by short-pulse signals. Function of sequention is realized by a digital device, named sequentor. The corresponding logical circuits are developed. They are represented in three forms according to the method used for decomposition of sequention. Sequentor contains venjunctors, which in turn are based on bistable cells. Such structure allows constructing regular models for sequential circuit design. The corresponding procedures are proposed for synthesis of asynchronous digital devices. Available structural features are demonstrated by an example.

4.1 Trigger Function

4.1.1 Definition and Properties

All kinds of presented above triggers (see Sect. 2.2) in their implementations exploit, as a rule, a logical constructions based on the following general formulae:

$$Z = X \vee \overline{X} \angle \overline{Y},\qquad(4.1.1)$$

$$\overline{Z} = Y \vee \overline{Y} \angle \overline{X}.\qquad(4.1.2)$$

These formal expressions define a trigger function [1].

4.1.1.1 Restrictive Terms

To identify a trigger function it is necessary to obey the following conditions:

- $X \neq 0,\ Y \neq 0$;
- $X \neq \overline{Y}$;

V. Vasyukevich: Asynchronous Operators of Sequential Logic, LNEE 101, pp. 77–123.
springerlink.com   © Springer-Verlag Berlin Heidelberg 2011

- $X \wedge Y = 0$;

- $\overline{X} \angle \overline{Y} \neq 0$;

- $\overline{Y} \angle \overline{X} \neq 0$.

In the case when the given requirements are wholly satisfied, any function is recognized as a trigger function, and a procedure of its realization by logical device of trigger type becomes much easier.

Trigger function is controlled by subfunctions X and Y as well as by venjunctions $\overline{X} \angle \overline{Y}$ and $\overline{Y} \angle \overline{X}$. Unity signal $X = 1$ forms the corresponding output state of $Z = 1$. Zero state is set by a signal $Y = 1$.

A role of keeper for fixed stable states is assigned to mirror venjunctions. Switching $X = 1/0$ on the background $Y = 0$ maintains the function in its unity state, and switching $Y = 1/0$ on the background $X = 0$ holds zero state.

Trigger function in essence is a specific representation form for venjunctive functions of asynchronous logic. This specificity is expressed by means of the following features.

4.1.2 Trigger Function Features

4.1.2.1 Negation without Calculations

A result of logical negation or inversion procedure for trigger function is a trigger function. Structure of the function does not change. Negation as a logical operation is reduced to interchanging the arguments: subfunction X is replaced with subfunction Y, and vice versa.

4.1.2.2 Conflict-Free Settings

Conflicts of settings are usually provoked by simultaneous actions of signals with opposite purposes: $X = 1$ and $Y = 1$. Trigger function is protected from suchlike conflicts by definition, owing to zero equality $X \wedge Y = 0$. Conflicting situation between output values $Z = 1$ and $Z = 0$ is expelled already at function level as it is seen from the conjunction, which is reduced to zero as follows:

$$Z \wedge \overline{Z} = ((X \vee \overline{X} \angle \overline{Y}) \wedge (Y \vee \overline{Y} \angle \overline{X})) = X \wedge Y) = 0 . \qquad (4.2)$$

But if inequality $X \wedge Y \neq 0$ is allowed, then $Z \wedge \overline{Z} \neq 0$ that is not logically correct at least in the framework of Boolean algebra.

4.1.2.3 Stabilization of Settings

Stabilization of set states is ensured in trigger function owing to inequalities $\overline{X} \angle \overline{Y} \neq 0$ and $\overline{Y} \angle \overline{X} \neq 0$. Corresponding venjunctions are not zero, and so each of them is able to keep its own state. Conflicts between the states are expelled, because of zero conjunction: $(\overline{X} \angle \overline{Y}) \wedge (\overline{Y} \angle \overline{X}) = 0$.

4.1.2.4 Realization Features

Trigger function is performed by realizing the corresponding operations at the outputs Z_X and Z_Y, logical unity values of which are set by signals X and Y respectively. Obeying the rule: $Z_X \wedge Z_Y = Z \wedge \overline{Z} = 0$, output signals are always in antiphase:

- if $Z_X = 1$, then $Z_Y = 0$; if $Z_X = 0$, then $Z_Y = 1$; and vice versa.

Trigger function ensures two stable output states [01] and [10] expressed in the format [$Z_X\ Z_Y$], where $Z_X = \overline{Z}_Y$. This mode of trigger functioning is called as paraphase.

As being paraphased, trigger is symmetrical if its setting functions are identical with accuracy to the variables interchanging.

4.1.3 Conditional Trigger Functions

If some function is structurally constructed in accordance with the trigger formulae, and along with this not wholly satisfies the required conditions, the function is called conditional trigger function. These functions are distinct because they obey the trigger logic at restricted set of input switchings. There are different reasons, for example, conflicts of settings, race hazards, functional independency of certain logical elements, and unacceptable oscillation of signals.

Conditional trigger functions are suitable for implementing the following logical devices.

1. Single phase (one output, two states) triggers.
2. Non-paraphase (two outputs, three states) triggers.
3. Paraphase (two outputs, two states) triggers that function in the mode of blocking.
4. Triggers with unacceptable signals and hazardous switchings.

Conditional trigger functions are wide spread in practice. In particular, suchlike functions are performed by bistable cell, SR flip-flop, static toggle flip-flop, asynchronous JK flip-flop.

4.1.4 Memory Formula

Memory formula is intended for representing such ability of trigger function that allows storing the set state. Memory formula accumulates binary sets and switchings, which do not entail setting actions. To be formally expressed, memory formula is represented by the function:

$$\Phi = \overline{X} \wedge \overline{Y}, \tag{4.3}$$

where trigger settings are negated.

For example, memory of SR flip-flop is expressed by the following formula:

$$\Phi = S \wedge R \vee \overline{S} \wedge \overline{R}, \tag{4.4}$$

according to which and taking into account the locked binary set $[SR] \neq [11]$, output state is stored during the time of $[SR] = [00]$. As for static D-latch, its memory is maintained by zero value of clock signal, as follows:

$$\Phi = \overline{(T \wedge D)} \wedge \overline{(T \wedge \overline{D})} = \overline{T}. \tag{4.5}$$

Memory formula is related with trigger function by means of the rule:

–   if $\Phi = 0/1$, then $Z = 1/1$ or $Z = 0/0$.

4.1.5 Enumeration of Trigger Functions

In the context of enumeration it is assumed that trigger function is given by formulae, which define settings X, Y and memory Φ in the form of compact expressions. These formulae consist of the following components: venjunctions and conjunctions, as operations of input variables x, y as well as variables themselves. The mentioned components are united by disjunctive operation. Thus, the problem of enumeration is reduced to distribution of the whole set of venjunctive functions among three nonempty and non-intersected subsets $\{X\}$, $\{Y\}$ and $\{\Phi\}$. Examples of the given distribution are represented below.

Example 1

Distribution, characteristic of static D-latch (see Sect. 2.2.4, Table 2.8a):

– $X = (x \angle y \vee y \angle x) = x \wedge y$;
– $Y = (x \angle \overline{y} \vee \overline{y} \angle x) = x \wedge \overline{y}$;
– $\Phi = (\overline{x} \angle y \vee \overline{x} \angle \overline{y} \vee y \angle \overline{x} \vee \overline{y} \angle \overline{x}) = \overline{x}$.

Example 2

Distribution, characteristic of dynamic D-latch (see Sect. 2.2.4, Table 2.8b):

– $X = x \angle y$;
– $Y = x \angle \overline{y}$;
– $\Phi = (\overline{x} \angle y \vee \overline{x} \angle \overline{y} \vee y \angle \overline{x} \vee \overline{y} \angle \overline{x} \vee y \angle x \vee \overline{y} \angle x) = \overline{x} \vee y \angle x \vee \overline{y} \angle x$.

Example 3

Distribution, characteristic of Muller's C-element (see Sect. 4.2.2, Table 4.3):

- $X = (x \angle y \vee y \angle x) = x \wedge y$;
- $Y = (\overline{x} \angle \overline{y} \vee \overline{y} \wedge \overline{x}) = \overline{x} \wedge \overline{y}$;
- $\Phi = (\overline{x} \angle y \vee x \angle \overline{y} \vee y \angle \overline{x} \vee \overline{y} \angle x) = x \oplus y$.

Initial group of elements, as forming components for trigger subsets, consists of eight venjunctions (see Sect. 1.8.1) without loss of generality. The whole number of ways to partition into three nonempty and non-intersected subsets is defined by the Stirling number of the second kind that means $S(8,3) = 966$. As applied to trigger functions this number is essentially smaller because of the restrictive terms (see Sect. 4.1.1). From the viewpoint of practice it is rational to consider only such partitions that are suitable for typical trigger distributions. Doing so it is important to show how many venjunctions, conjunctions and binary variables are distributed in each trigger subset.

The list of enumeration is represented in Appendix B. The corresponding data are formed as a table. The way of enumeration is based on the combinations of two subsets, namely {X} and {Y}. All possible (that means allowed) combinations of venjunctions, conjunctions and binary variables are demonstrated in Table 4.1 by square root "√".

Table 4.1 Combinability of components of setting functions.

	num-ber	Venjunctions				Conjunctions	Variables
		1	2	3	4	1	1
	1	√	√	√	√	√	√
Venjuntions	2	√	√	√	√	√	√
	3	√	√	√	√	√	-
	4	√	√	√	-	-	-
Conjunctions	1	√	√	√	-	√	√
Variables	1	√	√	-	-	√	-

Judging from table the following assertions take place.

1. Maximum number of combined venjunctions is 7 (3 and 4).
2. Maximum number of combined conjunctions is 2 (1 and 1).

3. Binary variables do not combine with each other.
4. Conjunctions do not combine only with four venjunctions.
5. Four venjunctions do not combine with Boolean components.
6. Three venjunctions do not combine only with binary variable.
7. Single venjunction and pair of venjunctions combine with a maximum number of possible components, namely six.

Note

Initially, trigger function was considered to be an analytical representation for trigger-type devices. However along with this, trigger function possesses fundamental opportunity to represent wide variety of venjunctive expressions on the basis of this specific form. In particular, it can be used as a sort of template for generating various logical devices. As well trigger function is able to represent sequential devices with memory that can not be displayed using venjunctive complete form.

4.2 Trigger-Type Devices

Trigger function is a mathematical model of varied trigger-type devices. Some of devices constructed on the basis of this model are demonstrated below. For the sake of definiteness asynchronous triggers of two variables x and y are explored.

4.2.1 Triggered Lambda-Function

Introduced above (see Sect. 1.6.4) auxiliary variable λ (*lambda*) is controlled by switchings $x = 0/1$ on the background $y =1$, and $y = 0/1$ on the background $x = 1$. In the context of trigger functions these switchings serve as a basis for defining setting subfunctions by the following venjunctions: $X = x \angle y$ and $Y = y \angle x$.

Availability of settings permits to characterize the logic of *lambda*-variable behavior by means of trigger type function:

$$Z = x \angle y \vee \overline{(x \angle y)} \angle \overline{(y \angle x)}. \tag{4.6}$$

It is real trigger function because all required conditions (see Sect. 4.1.1) are satisfied by the following expressions:

-     $X \neq \overline{Y}$ because $(x \angle y) \neq (\overline{x} \vee \overline{y} \vee x \angle y)$;
-     $X \wedge Y = 0$ because $(x \angle y) \wedge (y \angle x) = 0$;
-     $\overline{X} \angle \overline{Y} \neq 0$ (for example, $x = 0/1/0$ on the background $y = 1$);
-     $\overline{Y} \angle \overline{X} \neq 0$ (for example, $y = 0/1/0$ on the background $x = 1$).

Memory formula is formed as a result of the following calculations:

$$\Phi = (\overline{(x \angle y)} \wedge \overline{(y \angle x)}) = (\overline{x} \vee \overline{y} \vee y \angle x) \wedge (\overline{x} \vee \overline{y} \vee x \angle y)) = \overline{x} \vee \overline{y}. \tag{4.7}$$

Logic of *lambda*-trigger functioning is presented in Table 4.2.

Table 4.2 *Lambda*-trigger functioning.

		Y			
		1	0/1	1/0	0
	1	-	0	J	-
X	0/1	0	-	-	J'
	1/0	J	-	-	J'
	0	-	J'	J'	-

The corresponding graph is shown in Fig. 4.1.

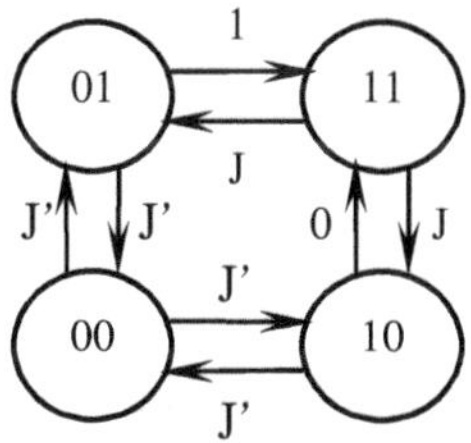

Fig. 4.1 Graph of *lambda*-function

In Fig. 4.2, a structural logic circuit for *lambda*-trigger is displayed. It is constructed on the basis of asynchronous elements: double venjunctor DV (see Sect. 2.3.2, Fig. 2.5) and negated bistable cell $\overline{BC}$ (see Sect. 2.1.2, Fig. 2.2). Two outputs of double venjunctor are linked to the inputs of negated bistable cell.

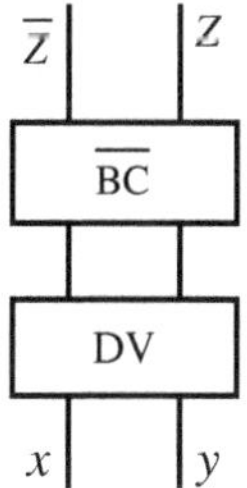

Fig. 4.2 Structural circuit of *lambda*-trigger

4.2.2 Trigger Function of C-Element (Muller C-Gate)

C-element (Muller C-gate) is asynchronous logical gate. Regardless of realization, the logic of its functioning is certainly presented by Table 4.3.

Tabulated data are adequately converted into the following venjunctive forms:

$$C = A \wedge B \vee (\overline{A} \angle B) \vee (\overline{B} \angle A), \qquad (4.8.1)$$

$$\overline{C} = \overline{A} \wedge \overline{B} \vee (A \angle \overline{B}) \vee (B \angle \overline{A}). \qquad (4.8.2)$$

Table 4.3 C-element functioning.

			B		
		1	0/1	1/0	0
	1	-	1	1	-
A	0/1	1	-	-	0
	1/0	1	-	-	0
	0	-	0	0	-

Output unity value is set by the binary combination $[AB] = [11]$, and zero value – by combination $[AB] = [00]$. Other switchings ensure a storage mode for output states.

Setting signals are formed by conjunctions $X = A \wedge B$ and $Y = \overline{A} \wedge \overline{B}$, which are wholly applicable for constructing a trigger function because of their satisfaction to the following rules (see Sect. 4.1.1):

- $X \neq \overline{Y}$ because $(A \wedge B) \neq (A \vee B))$;
- $X \wedge Y = 0$ because $(A \wedge B) \wedge (\overline{A} \wedge \overline{B}) = 0$;
- $\overline{X} \angle \overline{Y} \neq 0$ (for example, $A = 1/0$ on the background $B = 1$);
- $\overline{Y} \angle \overline{X} \neq 0$ (for example, $A = 0/1$ on the background $B = 0$).

Thus, C-element realizes trigger function as follows:

$$C = A \wedge B \vee (\overline{A \wedge B}) \angle (\overline{\overline{A} \wedge \overline{B}}). \qquad (4.9)$$

Memory formula combines input signals by means of XOR operation in accordance with the expression:

$$\Phi = (\overline{X} \wedge \overline{Y} = A \wedge \overline{B} \vee \overline{A} \wedge B) = A \oplus B. \qquad (4.10)$$

4.2.3 Trigger Function for Doubled Bistable Cell

In Fig. 4.3 a logical circuit is displayed. It consists of two bistable cells connected in such a manner that outputs of one bistable cell are attached to inputs of another cell.

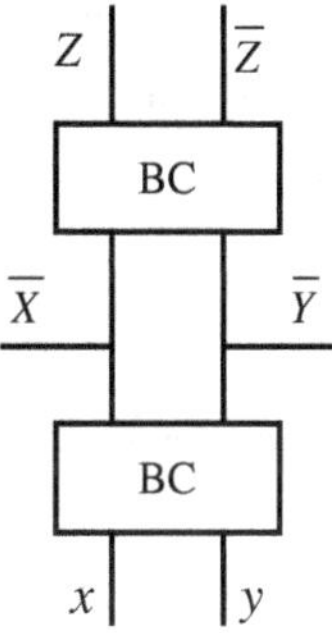

Fig. 4.3 Bistable cells connection

Function realized by the given circuit is defined by the following formulae:

$$Z = (\overline{\overline{x} \vee x \angle y}) \vee (\overline{x} \vee x \angle y) \angle (\overline{y} \vee y \angle x)\,, \tag{4.11.1}$$

$$\overline{Z} = (\overline{\overline{y} \vee y \angle x}) \vee (\overline{y} \vee y \angle x) \angle (\overline{x} \vee x \angle y)\,. \tag{4.11.2}$$

Required settings are caused by subfunctions:

$$-\quad X = (\overline{\overline{x} \vee x \angle y})\,;\; Y = (\overline{\overline{y} \vee y \angle x})\,.$$

The logic of functioning is presented by the corresponding truth table (Table 4.4), and switching graph (Fig. 4.4).

Table 4.4 Device (Fig. 4.3) functioning.

		|	y		
		1	0/1	1/0	0
x	1	-	1	1	-
	0/1	0	-	-	1
	1/0	0	-	-	1
	0	-	0	0	-

Conditions, required for the function of doubled bistable cell to be considered as trigger function (see Sect. 4.1.1), are satisfied in view of the expressions:

– $X \neq \overline{Y}$ because $(x \wedge \overline{y} \vee y \angle x) \neq (\overline{y} \vee y \angle x)$;

– $X \wedge Y = 0$ because $(x \wedge \overline{y} \vee y \angle x) \wedge (y \wedge \overline{x} \vee x \angle y)) = 0$

– $\overline{X} \angle \overline{Y} \neq 0$ (for example, $x = 1/0$ on the background $y = 0$);

– $\overline{Y} \angle \overline{X} \neq 0$ (for example, $y = 1/0$ on the background $x = 0$).

Memory of the doubled bistable cell is given by the formula: $\Phi = \overline{x} \wedge \overline{y}$.

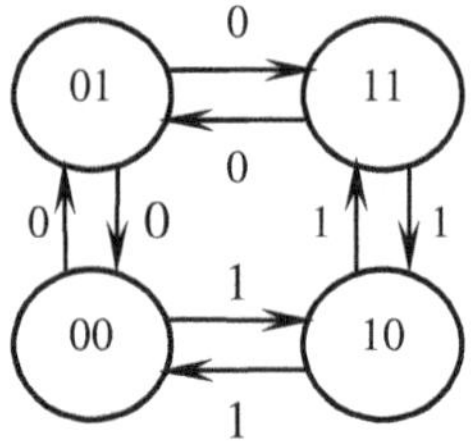

Fig. 4.4 Graphic image of the Table 4.4

Remark

As presented in Table 4.4, the logic of doubled cell functioning copies the similar logic of C-element (Table 4.3) with an accuracy to the input associations: $A = x$, $B = \overline{y}$.

4.2.4 Examples of Trigger Devices

Due to venjunction applicability for constructing sequential trigger-type circuits, a set of basic asynchronous elements and devices can be essentially expanded and varied. Area of the expansion is limited, because of the limits of venjunctive complete form itself.

In the context of trigger functions, logical devices with atypical, that means venjunctive settings, are of special interest; as well as devices, storage mode of which instead of usual binary combinations is maintained by input switchings. For example:

$$\{X, Y, \Phi\} = \{(\overline{x} \angle y \vee \overline{y} \angle \overline{x}), (x \wedge y \vee y \angle \overline{x}), (x \wedge \overline{y})\} ; \qquad (4.12.1)$$

$$\{X, Y, \Phi\} = \{(x \wedge y \vee \overline{y} \angle \overline{x}), (x \wedge \overline{y}), (\overline{x} \wedge y \vee \overline{x} \angle \overline{y})\} . \qquad (4.12.2)$$

Other examples accompanied by graphs of switchings are gathered in Appendix C. Selected devices are characterized by trigger functions with the following features:

1. Settings by means of venjunctions
2. Settings by means of a pair of mirror venjunctions

3. Settings by means of two pairs of mirror venjunctions
4. Settings by means of three pairs of mirror venjunctions
5. Memory formula, expressed with a single venjunction
6. Memory formulas, expressed with two venjunctions
7. Memory formulas, expressed with three venjunctions
8. Memory formulas, expressed with four venjunctions

4.2.5 Notes on Classification of Triggers

It is obvious that trigger function and results obtained on the basis of this function are able to influence upon the classification of trigger circuits. Owing to venjunction, new distinguishing features are produced and, as to traditional criterions, they are filled with new content.

According to existing classification, a trigger (flip-flop, latch) is considered to be static, if its settings are held by the corresponding signals while these signals are active. On the other hand in dynamic trigger all settings are performed at the moments of signals switchings, and therefore these settings do not need support.

Venjunction being dynamical operation is able to form dynamic mode for digital devices generally, and for trigger circuits particularly. In this context venjunctive function at the inputs is a natural indication that trigger circuit belongs to dynamic type.

Symmetry of trigger device is defined by settings X and Y. Corresponding logical formulae of trigger function must be invariant in relation to interchanging of variables. In the framework of binary algebra this condition can be satisfied with Boolean conjunctions and mirror venjunctions of two variables. Fore example:

- $\{X,Y\} = \{x \wedge \overline{y}, \overline{y} \wedge x\}$;
- $\{X,Y\} = \{x \angle y, y \angle x\}$;
- $\{X,Y\} = \{x \wedge \overline{y} \vee x \diagup y, \overline{x} \wedge y \vee y \diagup x\}$.

Aside from the input features a memory formula is also able to vary classes of triggers. On the one hand this formula can be expressed in its Boolean form, that is, without venjunctions, and therefore named combinational. On the other hand, memory formula can be represented in venjunctive, that is, non-combinational form. Combinational memory is characteristic of static triggers and symmetric triggers.

For clarification the following examples are supplied with additional distinguishers of the proposed classification.

A symmetric static trigger with combinational memory:

$$X = x \wedge \overline{y}, \ Y = \overline{x} \wedge y, \ \Phi = x \wedge y \vee \overline{x} \wedge \overline{y}. \tag{4.13.1}$$

An asymmetric static trigger with combinational memory:

$$X = x, \ Y = \overline{x} \wedge y, \ \Phi = \overline{x} \wedge \overline{y}. \tag{4.13.2}$$

A symmetric dynamic trigger with combinational memory:

$$X = x \angle y, \quad Y = y \angle x, \quad \Phi = \overline{x} \vee \overline{y}. \qquad (4.13.3)$$

An asymmetric trigger with non-combinational memory:

$$X = x \angle y, \quad Y = x \wedge \overline{y}, \quad \Phi = \overline{x} \vee y \angle x. \qquad (4.13.4)$$

An asymmetric dynamic trigger with non-combinational memory:

$$X = x \angle y, \quad Y = \overline{x} \wedge y \vee \overline{x} \angle \overline{y}, \quad \Phi = x \wedge \overline{y} \vee y \angle x \vee \overline{y} \angle \overline{x}. \qquad (4.13.5)$$

4.3 Zone Model for Switching Functions

Trigger-type switching function always stays in one of three states. They are: the state of unity setting, the state of zero setting, and the state of memory. In accordance with mentioned states, three zones are formed (Fig. 4.5).

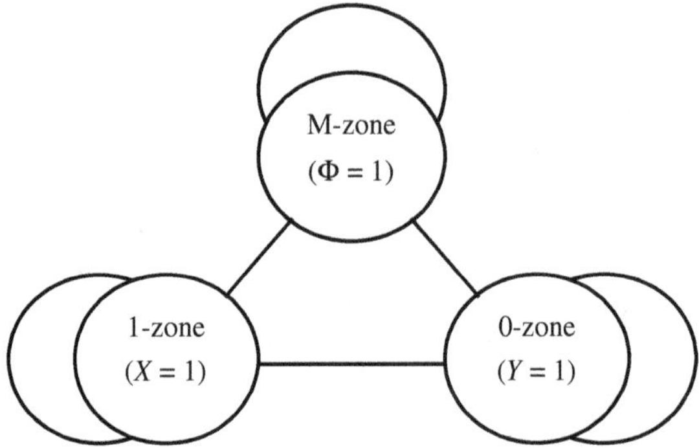

Fig. 4.5 Zone model for switching function

1-zone is intended for logical unity settings. It unites all input signals (binary sets and their sequences) required for output unity value setting. Within this zone, setting signals of X subfunction are active. While and because signal $X = 1$ acts, trigger device holds the unity state $Q = 1$.

0-zone is intended for logical zero settings. It unites all input signals (binary sets and their sequences) required for output zero value setting. Within this zone, setting signals of Y subfunction are active. While and because signal $Y = 1$ acts, trigger device holds the zero state $Q = 0$.

M-zone is intended for storing the output state that has been set previously. It unites all input signals (binary sets and their sequences) required for keeping output state in its unity or zero value. Within this zone, memory formula is active. While and because signal $\Phi = 1$ acts, trigger device holds the constant output value, invariability of which is maintained due to pseudoswitchings $Q = 1/1$ and $Q = 0/0$.

All three zones are connected together by transition channels. Transition from 0-zone into 1-zone is initiated by switching $X = 0/1$ on the background $\Phi = 0$, and is accompanied by zeroing $Y = 1/0$ with setting $Q = 0/1$. Reverse transition from 1-zone into 0-zone is initiated by switching $Y = 0/1$ on the background $\Phi = 0$, and is accompanied by zeroing $X = 1/0$ with setting $Q = 1/0$.

Transition from 1-zone into M-zone is initiated by switching $X = 1/0$ on the background $Y = 0$, and is accompanied by switching $\Phi = 0/1$ with keeping $Q = 1/1$. Transition from 0-zone into M-zone is initiated by switching $Y = 1/0$ on the background $X = 0$, and is accompanied by switching $\Phi = 0/1$ with keeping $Q = 0/0$.

Transition from M-zone into 1-zone is initiated by switching $X = 0/1$ on the background $Y = 0$, and is accompanied by zeroing $\Phi = 1/0$ with setting $Q = 1$. Transition from M-zone into 0-zone is initiated by switching $Y = 0/1$ on the background $X = 0$, and is accompanied by zeroing $\Phi = 1/0$ with setting $Q = 0$.

There are transitions, which do not presume moving to other zone. They are called intrazone (as contrast with interzone) transitions. Intrazone transitions are caused by signal actions, that occurred inside of zone and do not initiate exit from it. As a result, zone model retains its state. Switching function, subfunctions X and Y, and memory formula keep their states.

In accordance with the represented model, every switching of any input signal activates one of transitions, thus initiating the movement inside or outside the corresponding zone. Transition from one zone to another zone is performed without participation of the third zone, which holds zero state. Every transition process is completed by stable state setting.

Depending on the modeled function, some switchings can prove to be not realized. For this reason it is possible that some transitions, as well as their channels, "disappear". This circumstance is reflected in the objects of asynchronous logic. In the specific zone configurations only real (active) transitions are displayed.

4.3.1 Typical Zone Models of Trigger Circuits

Table 4.5 presents typical zone models of trigger-type circuits. Offered examples of eight models are constructed respectively to the eight examples of trigger functions which are gathered in Appendix C and used as sources for modeling.

Table 4.5 Typical zone models.

Example 1

	1-zone	0-zone	M-zone
1-zone	√		√
0-zone			√
M-zone	√	√	√

Example 2

	1-zone	0-zone	M-zone
1-zone			√
0-zone			√
M-zone	√	√	√

Table 4.5 (*continued*)

Example 3

⟶↑	1-zone	0-zone	M-zone
1-zone			√
0-zone			√
M-zone	√	√	

Example 4

⟶↑	1-zone	0-zone	M-zone
1-zone	√	√	√
0-zone	√	√	√
M-zone	√	√	

Example 5

⟶↑	1-zone	0-zone	M-zone
1-zone	√	√	√
0-zone	√	√	√
M-zone	√		

Example 6

⟶↑	1-zone	0-zone	M-zone
1-zone	√	√	√
0-zone		√	√
M-zone	√		

Example 7

⟶↑	1-zone	0-zone	M-zone
1-zone		√	√
0-zone	√	√	√
M-zone	√	√	√

Example 8

⟶↑	1-zone	0-zone	M-zone
1-zone		√	√
0-zone	√		√
M-zone	√	√	√

Transition tables are structured as follows. Columns and rows are headed by 1-zone, 0-zone and M-zone. For definiteness sake, transitions are directed from vertically located zones to horizontally located zones that are shown by arrow. Every available transition is marked by a sign √.

As judged from example 1 the corresponding model has bi-directional transitions between memory zone (M-zone) on the one hand, and setting zones (1-zone and 0-zone) on the other hand. There are also internal transitions within a memory zone (M-zone) and a logical unity setting 1-zone. At the same time setting 0-zone is devoid of internal transitions. Interzone transitions which could connect 0-zone and 1-zone are absent as well.

In zone models represented by examples 2 and 3 there are no transitions between setting zones, and internal transition takes place only in example 2. Zone model, which is built for a source function of example 2, is also characteristic of dynamic D-latch (see Sect. 2.2.4, Table 2.8b), *lambda*-trigger (see Sect. 4.2.1, Table 4.2) and C-element (see Sect. 4.2.2, Table 4.3).

Zone model represented by example 3 does not contain intrazone transitions at all. This model has only four interzone transitions; they are used for connection of both setting zones with memory zone. Zone model, which is built for a source function of example 3, is also characteristic of SR flip-flop (see Sect. 2.2.1, Table 2.4).

Zone model represented by example 4 contains all transitions excluding internal one within memory zone. Aside from a source function such configuration is also characteristic of JK flip-flop (see Sect. 2.2.2, Eq. 2.14) and doubled bistable cell (see Sect. 4.2.3, Table 4.4).

In examples 5 and 6 a transition from memory zone into setting 0-zone is absent. Further to it, in the model represented by example 6 a transition from 0-zone into setting 1-zone is not in use.

Zone model represented by example 7 contains all transitions excluding internal one within setting 1-zone. Model represented by example 8 does not use intrazone transitions of the setting zones. Analogous configuration is also characteristic of static D-latch (see Sect. 2.2.4, Table 2.8a)

4.3.2 Basic Zone Configurations

Trigger-type devises are able to function under condition when the number of transitions is restricted. According to the trigger function definition (see Sect. 4.1.1), the following transitions must be obligatory used:

- both transitions from setting zones into the memory zone;
- at least one transition into setting 1-zone as well as into setting 0-zone.

Four transitions is a minimal number of interzone transitions for any model of trigger function. Corresponding basic configurations are presented in Fig. 4.6. Basic models are shown without optional intrazone transitions.

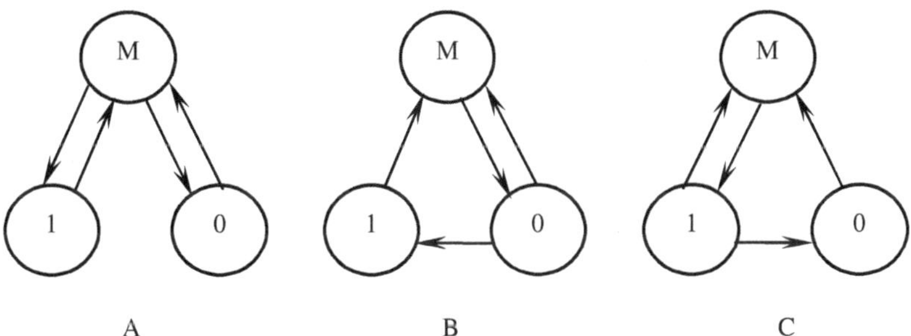

Fig. 4.6 Basic transitions of zone model

Typical zone configurations represented by the corresponding examples in Table 4.5, are connected with the basic models in the following manner.

Zone model represented by example 3 in Table 4.5 exactly conforms to the basic model A, which in respect of interzone transitions complies with examples 1 and 2. Zone models represented by examples 4, 7 and 8 are not restricted in interzone transitions. In contrast with the basic models these examples use not four, but all six possible transitions. Similar configuration is formed when basic model B is combined with basic model C that means "B plus C". Zone model represented by example 6 follows basic model C, which in the case of additional transition from 0-zone into 1-zone covers all five interzone transitions of example 5. Five interzone transitions are also formed in configurations which combine basic models using compositions "A plus B" and "A plus C".

4.3.2.1 Zone Model Symmetry

Zone configurations conformed to basic model A, as well as compositions formed by combining basic models B and C, are symmetric from the memory zone standpoint. This symmetry is adequately revealed in example tables (Table 4.5), if one's attention is drawn to the interzone transitions which are located about the diagonal that crosses squares assigned for intrazone transitions. In this context zone models, represented by examples 2, 3, 4 and 8 are symmetric models. Other zone models are considered to be asymmetrical.

In a general case the following configurations are asymmetric models.

1. Basic models B and C
2. Compositions "A plus B" and "A plus C"
3. Basic models supplemented with a transition between setting zones
4. Basic models B and C, supplemented with a transition from memory zone to setting 1-zone and 0-zone respectively.

4.3.3 *Example of Using Zone Model*

Asymmetrical zone model configured in accordance with the basic scheme B (Fig. 4.6).

Unity setting: $X = x \wedge y \vee \overline{x} \angle y \vee \overline{y} \angle x$.

Zero setting: $Y = x \angle \overline{y} \vee y \angle \overline{x}$.

Function: $Z = (x \wedge y \vee \overline{x} \angle y \vee \overline{y} \angle x) \vee \overline{(x \wedge y \vee \overline{x} \angle y \vee \overline{y} \angle x)} \angle \overline{(x \angle \overline{y} \vee y \angle \overline{x})}$.

Memory formula: $\Phi = \overline{x} \wedge \overline{y}$.

Cycles of switchings:

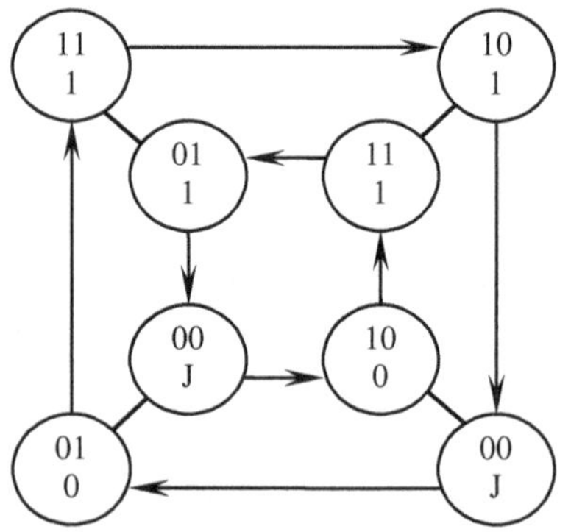

Zone model:

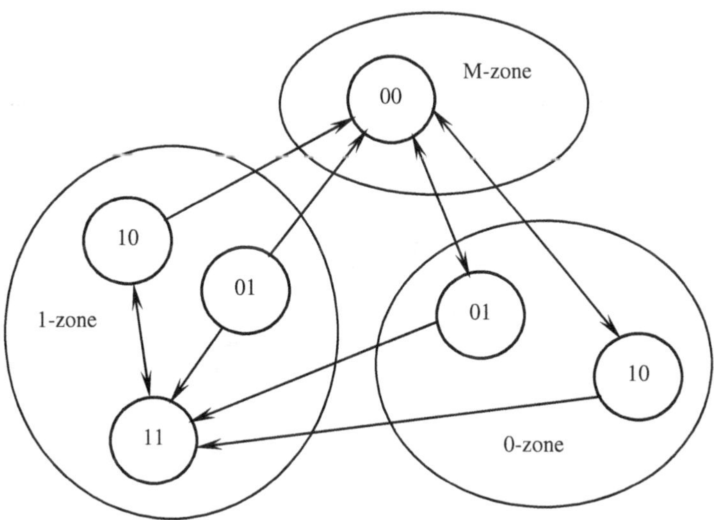

Graph of switchings:

4.3.4   Zone Model in the Context of Race Hazards

Transitions may come into conflict with each other because of signals race. Conflict is considered to be hazardous, if it happens between transitions from 1-zone (0-zone) into memory zone on the one hand, and into 0-zone (1-zone) on the other hand. Valid transition into M-zone holds the previous output state. Otherwise, invalid transition into setting zone changes output signal to opposite value. To avoid such a situation it is necessary to forbid hazardous switchings. Other way – to give preference to conflict-free zone models: basic model (A) in Fig. 4.6, exemplifying models (examples 1, 2, and 3) in Tables 4.5. As a result the conflict condition proves to be closed a priori, because one of the conflicting switchings loses its own transition.

4.4 Asynchronous Logic of Feedbacks

When constructing logical circuits intended for different purposes, quite often so called feedbacks are used. Feedback attaches the output of device to the input (or inputs) of the same device. So output signal becomes dependent on its own previous value, aside from usual input dependency. Existing feedbacks are subdivided into positive and negative. Their difference reveals itself in the event of essential self-dependency when input signals actually influence the returning output signal. If output signal returns in its initial value, the feedback is considered to be positive. Otherwise if output signal returns having opposite value, the feedback is negative.

4.4.1 Positive Feedback

When positive feedback is activated its output signal is identified by the binary dependency $Q = F(\ldots Q)$, in consequence of which the following logical options are presumed.

4.4.1.1 Feedback Is Given by the Formula $Q = Q$

It is feedback of closed-loop kind. Signal Q is permanently kept one of the constant values: 0 or 1. This situation can be considered as a re-circulated memorizing. The corresponding zone model is presented by active M-zone, outputs from which are not provided. Moreover, setting zones themselves are not provided too. As for the function, it is setting-free, and therefore its value is indeterminate: $Q = J$. Feedback function is defined by the equivalent relationship:

$$Q \Leftrightarrow 1 \angle 1, \tag{4.14}$$

where factor of indeterminacy is expressed by venjunction.

4.4.1.2 Feedback Is Given by the Formula $Q = x \wedge Q$

If $x = 0$, then $Q = 0$; but if $x = 1$, then $Q = Q$. From the view point of zone model, equality $Q = Q$ predestinates the presence of memory zone. Zero value $x = 0$ is the setting signal. It places the function into 0-zone. Following operations are reduced to transitions between zero and memory zones. Moving from 0-zone into M-zone is initiated by the switching $x = 0/1$. Under the action of the reverse switching $x = 1/0$, the function returns into the 0-zone. As for the feedback function definition, the following equivalent relation takes place:

$$(x \wedge Q) \Leftrightarrow (1 \angle x) . \tag{4.15}$$

The feature of equivalency remains valid after the logical negation is applied as follows:

$$\overline{(x \wedge Q)} = (\overline{x} \vee \overline{Q}) \Leftrightarrow \overline{(1 \angle x)} = (\overline{x} \vee x \angle 1) . \tag{4.16}$$

4.4.1.3 Feedback Is Given by the Formula $Q = x \vee Q$

If $x = 1$, then $Q = 1$; but if $x = 0$, then $Q = Q$. Signal $x = 1$ sets the output at the state of the logical unity. The corresponding zone model permits only the following transitions: from 1-zone into M-zone under the action of the switching $x = 1/0$, and back into the 1-zone under the action of the switching $x = 0/1$. In this case, the equivalent relationships are formed by involving the following functions:

$$(x \vee Q) \Leftrightarrow (x \vee \overline{x} \angle 1) , \qquad (4.17.1)$$

$$(\overline{x} \wedge \overline{Q}) \Leftrightarrow (1 \angle \overline{x}) . \qquad (4.17.2)$$

4.4.1.4 Feedback Is Given by the Formula $Q = x \angle Q$

If $x = 0$, then $Q = 0$; but if $x = 0/1$ on the background $Q = 1$, then $Q = Q$. Setting signal $x = 0$ places the function into the 0-zone. Further switchings maintain the intrazone transitions only, because it is impossible to set background value $Q = 1$ required for entering into 1-zone by switching $x = 0/1$. In spite of the fact that this zone model and the model considered above (see Sect. 4.4.1.2) are distinct from each other, the corresponding feedbacks are similar as to logical behavior. And therefore the feedback functions are equivalent:

$$(x \angle Q) \Leftrightarrow (x \wedge Q) , \qquad (4.18.1)$$

$$(\overline{x} \vee \overline{Q} \vee Q \angle x) \Leftrightarrow (\overline{x} \vee \overline{Q}) . \qquad (4.18.2)$$

4.4.1.5 Feedback Is Given by the Formula $Q = Q \angle x$

The signal $x = 0$ places the function into 0-zone, output from which is blocked because it is impossible to realize switching $Q = 0/1$ on the background $x = 1$. Switching $x = 0/1$ initiates intrazone transition of the 0-zone. As equivalent are considered to be the following functions:

$$(Q \angle x) \Leftrightarrow (x \wedge Q) , \qquad (4.19.1)$$

$$(\overline{x} \vee \overline{Q} \vee x \angle Q) \Leftrightarrow (\overline{x} \vee \overline{Q}) . \qquad (4.19.2)$$

4.4.1.6 Feedback Is Given by the Formula $Q = x \vee y \wedge Q$

This function belongs to the set of trigger kind functions. To be more precise, it is a conditional trigger function. The corresponding zone model contains all interzone transitions. Unity value $x = 1$ is the setting signal for 1-zone. Transition into 0-zone is ensured by the combination of signals $x = 0$ and $y = 0$. Equality $Q = Q$ required for memory zone is maintained by the combination of signals $x = 0$ and $y = 1$. As to behavior this feedback logically copies the bistable cell (Fig. 2.1) with an accuracy to renaming the signals X, Y and Z_X so that $\overline{X} = x$, $Y = y$, $Z_X = Q$. Thus, the following equivalent relationships are formed:

$$(x \vee y \wedge Q) \Leftrightarrow (x \vee \overline{x} \angle y), \tag{4.20.1}$$

$$(\overline{x} \wedge (\overline{y} \vee \overline{Q})) \Leftrightarrow (\overline{x} \wedge (\overline{y} \vee y \angle \overline{x})). \tag{4.20.2}$$

4.4.1.7 Feedback is Given by the Formula $Q = x \vee y \angle Q$

It is not trigger function. The feedback does not intend transitions for entering the memory zone. Unity state is caused by the signal $x = 1$ as well as by switching $y = 0/1$ on the background $Q = 1$. Other signals and switchings set zero value $Q = 0$. The following functions are equivalently related:

$$(x \vee y \angle Q) \Leftrightarrow (x \vee \overline{(y \angle x)} \angle y), \tag{4.21.1}$$

$$(\overline{x} \wedge (\overline{y} \vee \overline{Q} \vee (Q \angle y))) \Leftrightarrow (\overline{x} \wedge (\overline{y} \vee y \angle \overline{(y \angle x)})). \tag{4.21.2}$$

4.4.1.8 Feedback Is Given by the Formula $Q = x \vee Q \angle y$

It is not a trigger function. If $x = 1$ or $Q = 0/1$ on the background $y = 1$, then $Q = 1$. Other signals and switchings produce $Q = 0$. The following relationships are valid:

$$(x \vee Q \angle y) \Leftrightarrow (x \vee \overline{(x \angle y)} \angle y), \tag{4.22.1}$$

$$(\overline{x} \wedge (\overline{y} \vee \overline{Q} \vee (y \angle Q))) \Leftrightarrow (\overline{x} \wedge (\overline{y} \vee y \angle \overline{(x \angle y)})). \tag{4.22.2}$$

4.4.2 *Negative Feedback*

When negative feedback is activated its output signal is identified by the binary dependence of the following kind: $Q = F(\ldots \overline{Q})$. Output inversion causes functional collisions, which are not characteristic of asynchronous logic itself. This is reflected in the fact that protracted self-dependence causes the process of reverse signal oscillation, as well as a short-term self-dependence is revealed in the form of single impulse.

4.4.2.1 Oscillation and Its Interruption

The feedback is given by the formula $Q = x \wedge \overline{Q}$. Signal $z = 0$ ensures zero setting $Q = 0$. Signal $x = 1$ transforms the function into equality $Q = \overline{Q}$ that is characteristic of the closed negative feedback. Output signal Q is found to be in logically unstable condition because the switching 0/1 permanently alternates with the switching 1/0. In accordance with zone model, the function circulates between 0-zone and 1-zone so that the corresponding transitions are permanently active. This operating mode is related to oscillation that is considered to be unproductive within the framework of asynchronous logic. The output state is not stable, reversible signal does not fix its own value in the form of logical 1 or 0.

Alternated switchings $Q = 0/1$ and $Q = 1/0$ of the feedback can be used for interrupting the process of oscillation, and in this way for performing the required output value setting. The given feedback is provided by its zero setting, and therefore it is necessary to maintain the unity setting. Otherwise the function remains not completely defined, and will be considered as not full-fledged. Oscillating process needs to be interrupted at the time when signal $Q = 1$ acts. For this purpose the switching $Q = 0/1$ on the background $x = 1$ is suitable, for example, in the following form:

$$Q = x \wedge \overline{Q} \vee (Q \angle x).\tag{4.23}$$

Additional venjunction by its own signal $Q \angle x = 1$ stimulates the output setting at logical unity value, which remains constant until the moment of the switching $x = 1/0$. However in this case, the required output setting is not ensured, because the simultaneously occurred switching $(x \wedge \overline{Q}) - 1/0$ causes the undesirable race hazards. Preferable way to avoid the risk of invalid operation is provided by trigger functions. The problem can be solved by means of the following expressions:

$$Q = x \wedge (\overline{Q} \vee Q \angle x),\tag{4.24.1}$$

$$Q = \overline{(\overline{x} \vee x \angle Q)},\tag{4.24.2}$$

$$Q = (x \wedge \overline{Q}) \vee \overline{(x \wedge \overline{Q})} \angle x,\tag{4.24.3}$$

which are formed on the basis of the trigger function (see Sect. 4.1).

4.4.2.2 Retention of Short-Pulsed Setting

The feedback is given by the formula $Q = x \angle \overline{Q}$. If $x = 0$, then $Q = 0$. Signal $x = 1$ causes the output switching $Q = 0/1$. Then acting around the loop, the obtained unity signal produces the switching $(x \angle \overline{Q}) = 1/0$, owing to which the output signal returns to the initial zero state. So, the feedback instead of stable unity value generates a short-time pulse, which is causes by the switching $x = 0/1$ on the background $Q = 0$. In other words, the feedback generates imperfect function. In order for output setting to be really performed, a short-time pulse should be kept.

To retain setting – means to prolong short pulse, and in this way do not allow the function to be reduced to zero at the moment of switching $(x \angle \overline{Q}) = 1/0$. This operation is performed by entering the venjunctive component, due to which the feedback is expressed in a trigger-like form:

$$Q = x \angle \overline{Q} \vee \overline{(x \angle \overline{Q})} \angle x.\tag{4.25}$$

4.5 Algorithms for Extension of Sequentions

First of all it is assumed that extension of sequentions is performed by incremental (step by step) adding of elements. Every k-step is assigned for transforming (k-1)-sequention into (k)-sequention. Such extension in essence is composite procedure in contrast to decomposition as an opposite action. Therefore it is reasonable that composition is based on the same formulae, which were assigned for decomposition purposes (see Sect. 3.8), but with consideration that extension of sequentions is made in reverse order.

4.5.1 Algorithm I

Algorithm is based on the conjunctive decomposition of sequention (see Sect. 3.9.2).

$Step$ 1: $x_1 = \langle x_1 \rangle$;

$Step$ 2: $x_2 \angle \langle x_1 \rangle = \langle x_1\, x_2 \rangle$;

$Step$ 3: $\langle x_1\, x_2 \rangle \wedge \langle x_2\, x_3 \rangle = \langle x_1\, x_2\, x_3 \rangle$;

$Step$ 4: $\langle x_1\, x_2\, x_3 \rangle \wedge \langle x_3\, x_4 \rangle = \langle x_1\, x_2\, x_3\, x_4 \rangle$;

. . . .

$Step$ n: $\langle x_1\, x_2\, x_3\, x_4 \ldots x_{n-1} \rangle \wedge \langle x_{n-1}\, x_n \rangle = \langle x_1\, x_2\, x_3\, x_4 \ldots x_{n-1}\, x_n \rangle$.

4.5.2 Algorithm II

This algorithm is mirror with reference to the previous Algorithm I.

$Step$ 1: $\langle x_n \rangle = x_n$;

$Step$ 2: $x_n \angle \langle x_{n-1} \rangle = \langle x_{n-1}\, x_n \rangle$;

$Step$ 3: $\langle x_{n-2}\, x_{n-1} \rangle \wedge \langle x_{n-1}\, x_n \rangle = \langle x_{n-2}\, x_{n-1}\, x_n \rangle$;

$Step$ 4: $\langle x_{n-3}\, x_{n-2} \rangle \wedge \langle x_{n-2}\, x_{n-1}\, x_n \rangle = \langle x_{n-3}\, x_{n-2}\, x_{n-1}\, x_n \rangle$;

. . . .

$Step$ n: $\langle x_1\, x_2 \rangle \wedge \langle x_2 \ldots x_{n-3}\, x_{n-2}\, x_{n-1}\, x_n \rangle = \langle x_1\, x_2 \ldots x_{n-3}\, x_{n-2}\, x_{n-1}\, x_n \rangle$.

4.5.3 Algorithm III

Algorithm is based on the venjunctive form of decomposed separation of elements (see Sect. 3.9.3).

Step 1: $x_1 = \langle x_1 \rangle$;

Step 2: $x_2 \angle \langle x_1 \rangle = \langle \langle x_1 \rangle x_2 \rangle = \langle x_1 \, x_2 \rangle$;

Step 3: $x_3 \angle \langle x_1 \, x_2 \rangle = \langle \langle x_1 \, x_2 \rangle x_3 \rangle = \langle x_1 \, x_2 \, x_3 \rangle$;

Step 4: $x_4 \angle \langle x_1 \, x_2 \, x_3 \rangle = \langle \langle x_1 \, x_2 \, x_3 \rangle x_4 \rangle = \langle x_1 \, x_2 \, x_3 \, x_4 \rangle$;

. . . .

Step n: $x_n \angle \langle x_1 \, x_2 \, x_3 \ldots x_{n-1} \rangle = \langle \langle x_1 \, x_2 \, x_3 \ldots x_{n-1} \rangle x_n \rangle = \langle x_1 \, x_2 \, x_3 \ldots x_{n-1} \, x_n \rangle$.

Note

Only systemized representative forms fit for producing step by step procedures, which cause an increase in size of sequentions.

4.6 Sequentor

Above mentioned algorithms serve as a basis for construction of sequentor – logical device of special kind assigned for realization of sequentional function. Sequentor (*decoder* as prototype in [2]) is considered as a basic element for various circuits designed in the frameworks of asynchronous sequential logic.

Implementation of Algorithm I is performed by means of the logical circuit presented in Fig. 4.8.

This circuit contains conjunctors (AND) and venjunctors V_1, V_2, V_3 ... V_{n-2}, V_{n-1}. Outputs of venjunctors are connected with inputs of conjunctors. External input x_1 is connected with a passive input of venjunctor V_1, and input x_n – with an active input of venjunctor V_{n-1}. All remaining inputs are furcated so that each of them is attached to active and passive inputs of neighboring venjunctors.

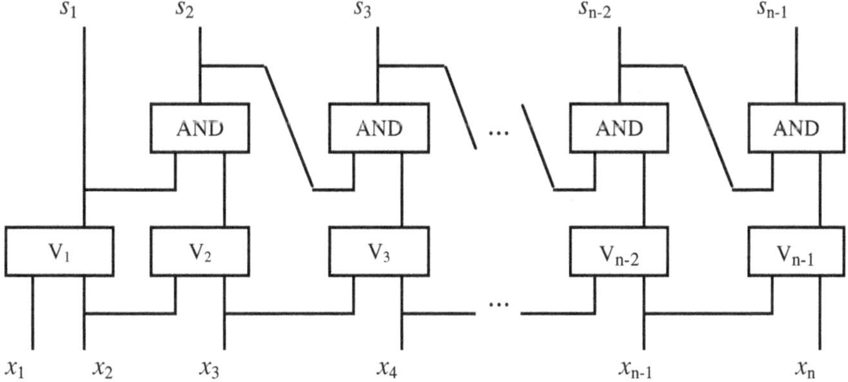

Fig. 4.8 Logical circuit of sequentor composed of venjunctors and conjunctors in accordance with the Algorithm I (see Sect. 4.5)

Outputs s_1, s_2, s_3 ... s_{n-2} and s_{n-1} are associated with the corresponding steps of the algorithm. By means of these outputs sequentional functions of two, three ... (n-1) and (n) variables are realized as follows:

- $\quad s_1 = \langle x_1\, x_2 \rangle$,
- $\quad s_2 = \langle x_1\, x_2\, x_3 \rangle$,
- $\quad s_3 = \langle x_1\, x_2\, x_3\, x_4 \rangle$,
- $\quad \ldots\ldots\ldots\ldots\ldots\ldots$
- $\quad s_{n-2} = \langle x_1\, x_2\, x_3\, x_4 \ldots x_{n-1} \rangle$,
- $\quad s_{n-1} = \langle x_1\, x_2\, x_3\, x_4 \ldots x_{n-1}\, x_n \rangle$.

Other variant of sequentor is represented in Fig. 4.9. Here Algorithm II (see Sect. 4.5) is chosen as a basis for construction of logical circuit. Structurally this circuit looks like the previous one (Fig. 4.8): output elements are conjunctors, whose inputs are connected with venjunctors. However in the case considered, sequentor resolves the problem differently as compared with Algorithm I.

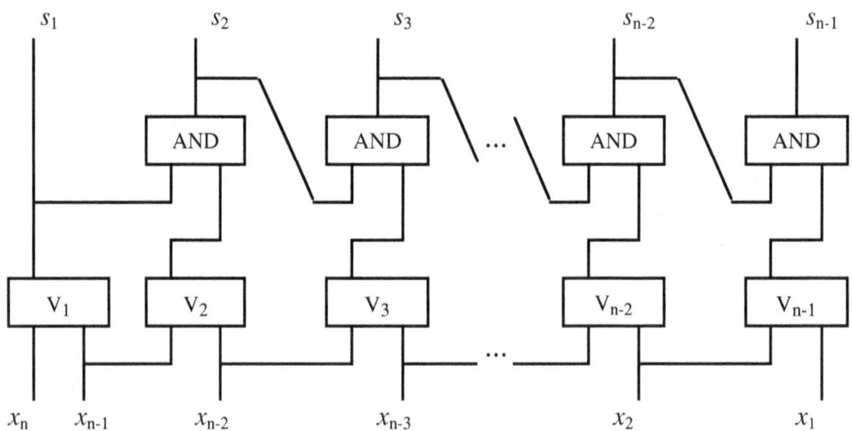

Fig. 4.9 Logical circuit of sequentor composed of venjunctors and conjunctors in accordance with the Algorithm II (see Sect. 4.5)

Outputs of the circuit realize the following functions:

- $\quad s_1 = \langle x_{n-1}\, x_n \rangle$,
- $\quad s_2 = \langle x_{n-2}\, x_{n-1}\, x_n \rangle$,
- $\quad s_3 = \langle x_{n-3}\, x_{n-2}\, x_{n-1}\, x_n \rangle$,
- $\quad \ldots\ldots\ldots\ldots\ldots\ldots$
- $\quad s_{n-2} = \langle x_2 \ldots x_{n-3}\, x_{n-2}\, x_{n-1}\, x_n \rangle$,
- $\quad s_{n-1} = \langle x_1\, x_2 \ldots x_{n-3}\, x_{n-2}\, x_{n-1}\, x_n \rangle$.

The third implementation of sequentor corresponds to Algorithm III (see Sect. 4.5). Logical circuit is represented in Fig. 4.10.

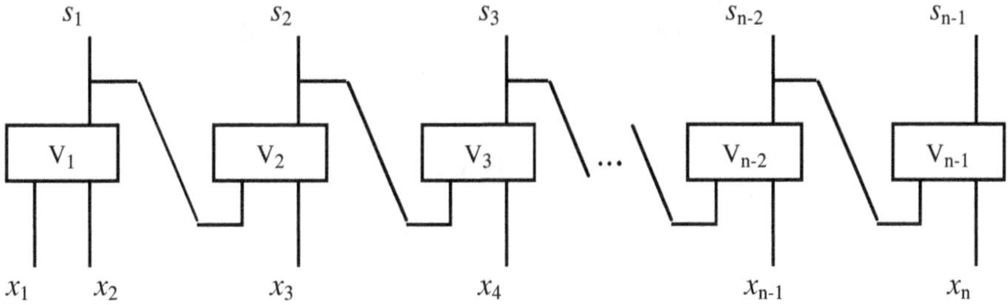

Fig. 4.10 Logical circuit of sequentor constructed as a cascade of venjunctors in accordance with the Algorithm III (see Sect. 4.5)

This circuit contains venjunctors and no conjunctors in contrast with the previous circuits (Fig. 4.8, 4.9). Venjunctors are connected consecutively (cascade like) so that output of each logical element except V_{n-1} is attached to the passive input of the next element. Passive input x_1 of venjunctor V_1, as well as active inputs of venjunctors V_1, V_2, V_3 ... V_{n-2}, V_{n-1}, are external inputs of the sequentor. Functions, which are realized on the outputs of the circuit, coincide with the analogous output functions of the device in Fig. 4.8.

4.7 Logical Implementation of Sequential Circuits

Historical Background

Since the middle of 1950s on the basis of finite automaton (Mealy machine [3] and Moore machine [4]) varied methods for sequential device synthesis were developed [5, 6]. Automaton was found to be an ideal mathematical model for synchronous sequential circuits, and rather acceptable model for asynchronous circuits [7, 8, 9].

At the same time it was observed that synchronous devices are restricted in speed, and asynchronous devices are encumbered with the race hazard problem. In this connection clockless (delay-insensitive, speed-independent and self-timed) circuits look more preferable because of their potentially high speed and hazardless behavior [10, 11].

For asynchronous circuits design finite-state machine, aperiodic automata [12], Petri net [13], NULL convention logic [14], and others mathematical apparatus are used.

4.7.1 Structural Model of Digital Devices

Developed above method of representing asynchronous memory of sequential circuits by means of venjunction and sequention operators, inevitably requires new architectural solutions in the field of digital circuit design. Generally on abstract level this solution is represented by four sections in Fig. 4.11.

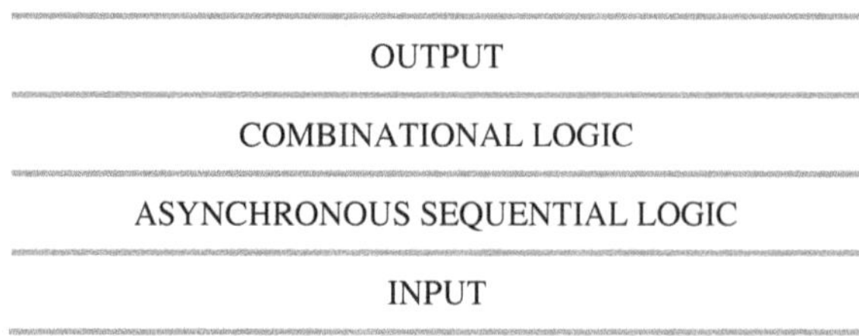

Fig. 4.11 Logic architecture of sequential circuit implementation

Input and output sections are empty of logic functions; they show a direction of logical transformations. Asynchronous sequential logic is required for realizing functions of memory. Memorized data enter the combinational section, where after logical processing output result is formed. Sequential and combinational implementations are located separately and consecutively so that logical memory is followed by logical combinations.

It is obvious that a block of asynchronous memory is the most difficult in realization. Additionally there are several useful variants. For these reasons it is rational to use theoretical base and developed above methods for representation of sequentions by means of conjunctive operators (see Sect. 3.8.2, 3.9.2, 3.11.1).

Operational elements needed for implementations of memory logic are represented in unrolled functional forms in Fig. 4.12. Structural components such as conjunctions, venjunctions, sequentions and elements of bistable memory are used.

Variant 1	Variant 2	Variant 3	Variant 4
Conjunctions	Conjunctions	Conjunctions	
Conjunctions	Conjunctions		Complicated sequentions
Conjunctions	Venjunctions	Elementary sequentions	
Bistable memory			

Fig. 4.12 Functional structures of logical memory

In accordance with a structuring way, memory block can be represented in four variants. They differ by a set of chosen components and their location. The following features are observed.

1. Bistable cell[1] together with above located conjunction (variant 1) on the one hand, and venjunction (variant 2) are functionally equivalent logic structures. The corresponding transformation is confirmed by the formula (see Sect. 2.3.1, Eq. 2.20).
2. Venjunctions together with above located conjunctions (variant 2) on the one hand, and elementary sequention (variant 3) on the other hand are functionally equivalent logic structures. The corresponding transformation is confirmed by the conjunctive representative form of sequention (see Sect. 3.9.2, Eq. 3.55).
3. Elementary sequentions together with above located conjunction (variant 3) on the one hand, and complicated sequention (variant 4) on the other hand are functionally equivalent logic structures. The corresponding transformation is confirmed by the conjunctive expansion (see Sect. 3.11.1, Eq. 3.66).

Operations given in Fig. 4.12 are realized by means of logical elements represented in Fig. 4.13. In contrast with number of functional structures (four variants) there are three methods of logical memory implementation. It is because complicated sequention can not be represented for general case; sequentions of these kinds are unlimited in embedding layers and combinations of elementary sequentions.

Method 1	Method 2	Method 3
Conjunctors	Conjunctors	
Conjunctors	Venjunctors	Sequentors
Bistable cells		
INPUT	INPUT	INPUT

Fig. 4.13 Three methods of logical memory implementation

Taking into account the variants in Fig. 4.12 and the methods in Fig. 4.13, abstract architecture of asynchronous logic implementation (Fig. 4.11) generates three structural forms available for constructing of sequential circuits. The corresponding models are displayed in Fig. 4.14. Properly speaking, every model gives a logic block diagram of the implemented circuit.

Sequential circuit is designed consecutively moving from input section to output section. Combinational circuit is built on the memory circuit. Combinational logic is represented in two-level disjunctive normal form so that conjunctors are followed by disjunctors

---

[1] Asynchronous memory of two states of one event with depth Md = 2 (see Sect. 3.12.2).

Model 1	Model 2	Model 3
OUTPUT	OUTPUT	OUTPUT
Disjunctor	Disjunctor	Disjunctor
Conjunctors	Conjunctors	Conjunctors
Bistable cells	Venjunctors	Sequentors
INPUT	INPUT	INPUT

Fig. 4.14 Structural models for asynchronous sequential circuit design

Remark

In spite of the fact that a certain sequential circuit can be designed in three ways, it is of little importance. From the functionality standpoint general structures are similar, even though the resulting circuits are found to be variously composed.

What implementation is better? It is obvious that the corresponding response does not fall into logic level; sooner it lies in the field of technical engineering and technology.

4.7.2 Advantages and Restrictions of the Model

As any theoretical findings, analytical methods based on the developed asynchronous operators have strong as well as weak aspects.

Advantages

1. Global feedbacks are excluded, and due to that hazard-free behavior of sequential device is ensured during races of switching signals.
2. Memory depth is not limited, and every required depth can be preassigned by means of enhancing the length of sequentors.
3. Procedure of sequential circuit design is implemented on the regular basis that concerns the order of actions as well as the set of asynchronous elements.

Restrictions

Proposed architecture is not intended for circuits whose behavior critically depends on:

-   internal, that is not input, signals (global feedbacks);
-   indeterminate signals (toggle flip-flop as example);
-   incorrect input sequences (see Sect. 3.2.3).

Note

In contrast with Fig. 4.9, sequentor in Fig. 4.10 is realized by means of venjunctors connected in a certain way; no conjunctors are in use. Therefore transitions from venjunctions to elementary sequentions (Fig. 4.12) as well as from venjunctors to sequentors (Fig. 4.13 and Fig. 4.14) are performed without supplemented conjunctions and conjunctors. Other elements of the displayed above structures remain in their places.

4.7.3 *Example of Constructing of Logical Circuit*

As an example, it is presumed to design a logical circuit for the following sequential function:

$$Q = y \wedge \langle x_1\, x_2\, x_3 \rangle \vee y \wedge \langle z_1\, z_2\, z_3\, z_4 \rangle \wedge \langle s_1\, s_2 \rangle \vee \langle\langle s_1\, s_2\, s_3\, s_4 \rangle \langle v_1\, v_2\, v_3 \rangle u_1\, u_2\, u_3\, u_4 \rangle$$

$$(4.26)$$

The function depends on the value of a binary variable y and certain sequences of values of other variables. In this connection, proposed in Fig. 4.15 realization has one output and nineteen inputs.

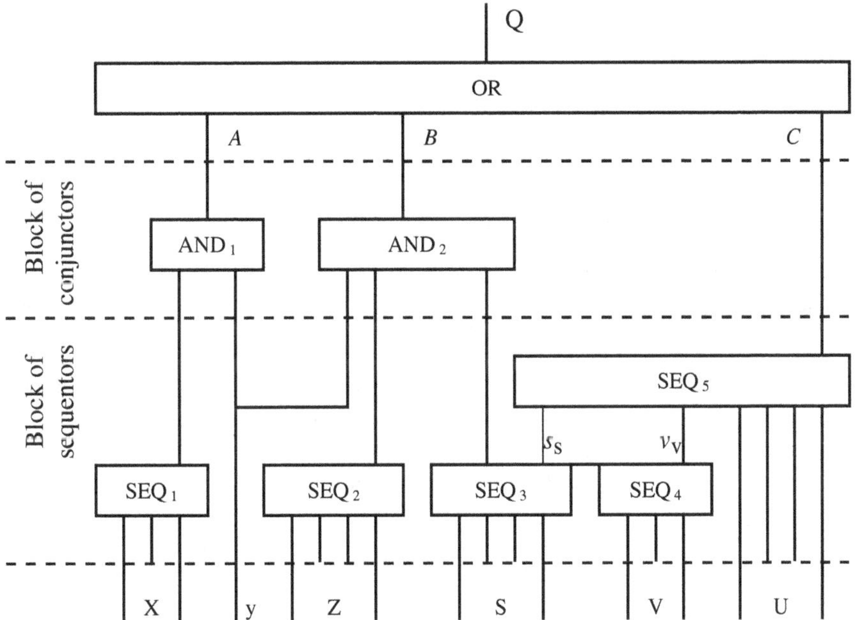

Fig. 4.15 Logical diagram for sequential function (Eq. 4.26)

All inputs, excluding one of them, are divided into five groups designated X, Z, S, V, U. Single input is marked y. Other inputs are distributed among groups as follows:

- x_1, x_2, and x_3 belong to X;
- z_1, z_2, z_3, and z_4 belong to Z;
- s_1, s_2, s_3, and s_4 belong to S;
- v_1, v_2, and v_3 belong to V;
- u_1, u_2, u_3, and u_4 belong to U.

In the example considered, construction of the corresponding logical diagram is performed on the basis of the structural model 3 (Fig. 4.14), which is developed in the previous Sect. 4.7.1.

Taking into account the mentioned model and implemented function (Eq. 4.26), input signals of groups X, Z, S and V enter the block of sequentors, including two three-input sequentors (SEQ_1 and SEQ_4) and two four-input sequentors (SEQ_2 and SEQ_3). At the outputs of the mentioned sequentors the following functions of elementary sequentions are realized:

- $\langle x_1\, x_2\, x_3 \rangle$ at output of SEQ_1;
- $\langle z_1\, z_2\, z_3\, z_4 \rangle$ at output of SEQ_2;
- $\langle s_1\, s_2 \rangle$ and $\langle s_1\, s_2\, s_3\, s_4 \rangle$ at output of SEQ_3;
- $\langle v_1\, v_2\, v_3 \rangle$ at output of SEQ_4.

Output of sequentor SEQ_3, which realizes a function of the sequention $\langle s_1\, s_2\, s_3\, s_4 \rangle$, and output of sequentor SEQ_4 are connected with input of sequention SEQ_5. Other inputs of this six-input sequentor receive signals from U group. Thus, at output of the sequentor a function of complicated sequention $\langle\langle s_1\, s_2\, s_3\, s_4 \rangle\langle v_1\, v_2\, v_3 \rangle u_1\, u_2\, u_3\, u_4 \rangle$ is realized.

Output of the sequentor SEQ_5 is directly attached to disjunctor. Output signals of other sequentors also enter disjunctor, but only after the corresponding logical transformation by a block of two conjunctors.

As a result of the enumerated above manipulations, at the output Q initial sequential function (Eq. 4.26) is realized.

According to the obtained circuit (Fig. 4.15) output signal is set at unity value $Q = 1$ in the case of $A = 1$ or $B = 1$, or $C = 1$, if the following conditions are met.

4.7.3.1 Conditions for $A = 1$

1. Signal value of y is logical unity: $y = 1$.
2. Signals of X group are switched as from $x_2 = 0/1$ on the background $x_1 = 1$ to finally $x_3 = 0/1$ on the background $x_2 = 1$.

4.7.3.2 Conditions for $B = 1$

1. Signal value of y is logical unity: $y = 1$.
2. Signals of Z group are switched as from $z_2 = 0/1$ on the background $z_1 = 1$, then $z_3 = 0/1$ on the background $z_2 = 1$ to finally $z_4 = 0/1$ on the background $z_3 = 1$.
3. Two first signals of S group are connected with a switching $s_2 = 0/1$ on the background $s_1 = 1$.

4.7.3.3 Conditions for $C = 1$

1. Signals of S group are switched as from $s_2 = 0/1$ on the background $s_1 = 1$, then $s_3 = 0/1$ on the background $s_2 = 1$, to finally $s_4 = 0/1$ on the background $s_3 = 1$.
2. Signals of S group are switched as from $v_2 = 0/1$ on the background $v_1 = 1$ to finally $v_3 = 0/1$ on the background $v_2 = 1$.
3. Output signals of the sequentors SEQ3 and SEQ4, as well as signals of U group are switched as follows: first $v_V = 0/1$ on the background $s_S = 1$, then $u_1 = 0/1$ on the background $v_V = 1$, $u_2 = 0/1$ on the background $u_1 = 1$, $u_3 = 0/1$ on the background $u_2 = 1$, to finally $u_4 = 0/1$ on the background $u_3 = 1$.

As seen from Fig. 4.14 (see Sect. 4.7.1), besides model 3 models 1 and 2 are also adequately suitable for sequential circuits design.

Logic diagrams obtained during the process of the function (4.26) realization, using the mentioned above structural models, are represented in Appendix D. They are more detailed (especially model 1) than the previously used model 3.

In Appendix D (Fig. D.1), model 2 is used. Implemented subfunctions are tentatively transformed by the rules of conjunctive expansion (see Sect. 3.11.1) as following:

$$A = y \wedge \langle x_1\, x_2 \rangle \wedge \langle x_2\, x_3 \rangle, \tag{4.27.1}$$

$$B = y \wedge \langle z_1\, z_2 \rangle \wedge \langle z_2\, z_3 \rangle \wedge \langle z_3\, z_4 \rangle \wedge \langle s_1\, s_2 \rangle, \tag{4.27.2}$$

$$C = \langle s_1\, s_2 \rangle \wedge \langle s_2\, s_3 \rangle \wedge \langle s_3\, s_4 \rangle \wedge \tag{4.27.3}$$
$$\langle v_1\, v_2 \rangle \wedge \langle v_2\, v_3 \rangle \wedge \langle u_1\, u_2 \rangle \wedge \langle u_2\, u_3 \rangle \wedge \langle u_3\, u_4 \rangle \wedge \langle s_4 v_3 \rangle \wedge \langle v_3\, u_1 \rangle$$

In Appendix D (Fig. D.2), model 1 is used. As an example, a fragment of the initial function is realized, namely the following subfunction:

$$C = \langle\langle s_1\, s_2\, s_3\, s_4 \rangle\langle v_1\, v_2\, v_3 \rangle u_1\, u_2\, u_3\, u_4 \rangle. \tag{4.28}$$

Bistable cells (BC) are built using NAND elements (see Sect. 2.1, Fig. 2.1). It is considered that venjunctive subfunctions are designed in accordance with a logical diagram in Fig. 2.3a (see Sect. 2.3.1).

4.7.3.4 Memory Characteristics

Memory depth (see Sect. 3.12.2) of the digital device constructed for function (4.26) is assigned by sequentors SEQ_3 and SEQ_5 (Fig. 4.15). This depth is defined by a length of the sequence $(s_1, s_2, s_3, s_4, v_3, u_1, u_2, u_3, u_4)$, and is equal to $Md = 9$.

Memory volume (see Sect. 3.12.4) is defined by a number of used venjunctors or bistable cells. Judging from the diagram (see App. D, Fig. D.1)) a total memory volume of the corresponding circuit is $Mv = 15$.

References

[1] Vasyukevich, V.: Analytics of trigger functions. ACCS Journal 43(4), 184–189 (2009)

[2] Vasyukevich, V.: Asynchronous sequences decoding. ACCS Journal 41(2), 93–99 (2007)

[3] Mealy, G.: Method to Synthesizing Sequential Circuits. Bell Systems Technical Journal, 1045–1079 (1955)

[4] Moore, E.: Gedanken-experiments on Sequential Machines. Automata Studies, Annals of Mathematical Studies 34, 129–153 (1956)

[5] Unger, S.: Asynchronous sequential switching circuits. Wiley Interscience, Hoboken (1969)

[6] Friedman, A., Menon, P.: Theory & design of switching circuits. Computer Science Press (1975)

[7] Birtwistle, G., Davis, A.: Asynchronous Digital Circuit Design. Springer, Heidelberg (1995)

[8] Brzozowski, J., Seger, C.-J.: Asynchronous Circuits (Monographs in Computer Science). Springer, Heidelberg (1995)

[9] Myers, C.: Asynchronous Circuit Design. Wiley Interscience, Hoboken (2001)

[10] Miller, R.: An Introduction to Speed Independent Circuit Theory. In: Proc. 2nd Ann. Symp. AIEE, pp. 85–110 (1961)

[11] Verhoeff, T.: Delay-insensitive codes – an overview. Distributed Computing 3(1), 1–8 (1988)

[12] Varshavsky, V., Astanovsky, A., et al.: Self-timed Control of Concurrent Processes: The Design of Aperiodic Logical Circuits in Computers and Discrete Systems. D. Raidel Publishing Company (1989)

[13] Petri, C.: Kommunikation mit Automaten. Institut für Instrumentelle Mathematik, Schriften des IIM, Nr. 2 (1962) (in German)

[14] Fant, K.: Logically determined design: clockless system design with NULL convention logic. A John Wiley & Sons, Chichester (2005)

Appendix A: The List of Venjunctive Functions on Two Variables

TRUNCATED FUNCTIONS		
Indefinite function (1)	Functions of one variable (2)	
$1 \angle 1$	$x \angle 1$	$1 \angle x$

EXCLUSIVELY VENJUNCTIVE FUNCTIONS							
Single venjunctions (8)							
$x \angle y$	$\bar{x} \angle \bar{y}$	$x \angle \bar{y}$	$\bar{x} \angle y$	$y \angle x$	$\bar{y} \angle \bar{x}$	$y \angle \bar{x}$	$\bar{y} \angle x$

Two venjunctions (24)			
$x \angle y \vee \bar{x} \angle \bar{y}$	$\bar{x} \angle \bar{y} \vee x \angle \bar{y}$	$x \angle \bar{y} \vee y \angle x$	$y \angle x \vee \bar{y} \angle \bar{x}$
$x \angle y \vee x \angle \bar{y}$	$\bar{x} \angle \bar{y} \vee \bar{x} \angle y$	$x \angle \bar{y} \vee \bar{y} \angle \bar{x}$	$y \angle x \vee y \angle \bar{x}$
$x \angle y \vee \bar{x} \angle y$	$\bar{x} \angle \bar{y} \vee y \angle x$	$x \angle \bar{y} \vee y \angle \bar{x}$	$y \angle x \vee \bar{y} \angle x$
$x \angle y \vee \bar{y} \angle \bar{x}$	$\bar{x} \angle \bar{y} \vee y \angle \bar{x}$	$\bar{x} \angle y \vee y \angle x$	$\bar{y} \angle \bar{x} \vee y \angle \bar{x}$
$x \angle y \vee y \angle \bar{x}$	$\bar{x} \angle \bar{y} \vee \bar{y} \angle x$	$\bar{x} \angle y \vee \bar{y} \angle \bar{x}$	$\bar{y} \angle \bar{x} \vee \bar{y} \angle x$
$x \angle y \vee \bar{y} \angle x$	$x \angle \bar{y} \vee \bar{x} \angle y$	$\bar{x} \angle y \vee \bar{y} \angle x$	$y \angle \bar{x} \vee \bar{y} \angle x$

Three venjunctions (32)		
$x \angle y \vee \bar{x} \angle \bar{y} \vee x \angle \bar{y}$	$x \angle y \vee \bar{x} \angle y \vee \bar{y} \angle x$	$\bar{x} \angle \bar{y} \vee \bar{x} \angle y \vee \bar{y} \angle x$
$x \angle \bar{y} \vee \bar{y} \angle \bar{x} \vee y \angle \bar{x}$	$x \angle y \vee \bar{x} \angle \bar{y} \vee \bar{x} \angle y$	$x \angle y \vee \bar{y} \angle \bar{x} \vee y \angle \bar{x}$
$\bar{x} \angle \bar{y} \vee y \angle x \vee y \angle \bar{x}$	$\bar{x} \angle y \vee y \angle x \vee \bar{y} \angle \bar{x}$	$x \angle y \vee \bar{x} \angle \bar{y} \vee y \angle \bar{x}$
$x \angle y \vee \bar{y} \angle \bar{x} \vee \bar{y} \angle x$	$\bar{x} \angle \bar{y} \vee y \angle x \vee \bar{y} \angle x$	$\bar{x} \angle y \vee y \angle x \vee \bar{y} \angle x$
$x \angle y \vee \bar{x} \angle \bar{y} \vee \bar{y} \angle x$	$x \angle y \vee y \angle \bar{x} \vee \bar{y} \angle x$	$\bar{x} \angle \bar{y} \vee y \angle \bar{x} \vee \bar{y} \angle x$
$\bar{x} \angle y \vee \bar{y} \angle \bar{x} \vee \bar{y} \angle x$	$x \angle y \vee x \angle \bar{y} \vee \bar{x} \angle y$	$\bar{x} \angle \bar{y} \vee x \angle \bar{y} \vee \bar{x} \angle y$
$x \angle \bar{y} \vee \bar{x} \angle y \vee y \angle x$	$y \angle x \vee \bar{y} \angle \bar{x} \vee y \angle \bar{x}$	$x \angle y \vee x \angle \bar{y} \vee \bar{y} \angle \bar{x}$
$\bar{x} \angle \bar{y} \vee x \angle \bar{y} \vee y \angle x$	$x \angle \bar{y} \vee \bar{x} \angle y \vee \bar{y} \angle \bar{x}$	$y \angle x \vee \bar{y} \angle \bar{x} \vee \bar{y} \angle x$

$x\angle y \vee x\angle\bar{y} \vee y\angle\bar{x}$	$\bar{x}\angle\bar{y} \vee x\angle\bar{y} \vee y\angle\bar{x}$	$x\angle\bar{y} \vee y\angle x \vee \bar{y}\angle\bar{x}$
$y\angle x \vee y\angle\bar{x} \vee \bar{y}\angle x$	$x\angle y \vee \bar{x}\angle y \vee \bar{y}\angle\bar{x}$	$\bar{x}\angle\bar{y} \vee \bar{x}\angle y \vee y\angle x$
$x\angle\bar{y} \vee y\angle x \vee y\angle\bar{x}$	$\bar{y}\angle\bar{x} \vee y\angle\bar{x} \vee \bar{y}\angle x$	-

Four venjunctions (16)

$x\angle y \vee \bar{x}\angle\bar{y} \vee x\angle\bar{y} \vee \bar{x}\angle y$	$\bar{x}\angle\bar{y} \vee x\angle\bar{y} \vee \bar{x}\angle y \vee y\angle x$
$x\angle y \vee \bar{x}\angle\bar{y} \vee x\angle\bar{y} \vee y\angle\bar{x}$	$\bar{x}\angle\bar{y} \vee x\angle\bar{y} \vee y\angle x \vee y\angle\bar{x}$
$x\angle y \vee \bar{x}\angle\bar{y} \vee \bar{x}\angle y \vee \bar{y}\angle x$	$\bar{x}\angle\bar{y} \vee \bar{x}\angle y \vee y\angle x \vee \bar{y}\angle x$
$x\angle y \vee x\angle\bar{y} \vee \bar{x}\angle y \vee \bar{y}\angle\bar{x}$	$\bar{x}\angle\bar{y} \vee y\angle x \vee y\angle\bar{x} \vee \bar{y}\angle x$
$x\angle y \vee \bar{x}\angle\bar{y} \vee y\angle\bar{x} \vee \bar{y}\angle x$	$x\angle\bar{y} \vee \bar{x}\angle y \vee y\angle x \vee \bar{y}\angle\bar{x}$
$x\angle y \vee x\angle\bar{y} \vee \bar{y}\angle\bar{x} \vee y\angle\bar{x}$	$x\angle\bar{y} \vee y\angle x \vee \bar{y}\angle\bar{x} \vee y\angle\bar{x}$
$x\angle y \vee \bar{x}\angle y \vee \bar{y}\angle\bar{x} \vee \bar{y}\angle x$	$\bar{x}\angle y \vee y\angle x \vee \bar{y}\angle\bar{x} \vee \bar{y}\angle\bar{x}$
$x\angle y \vee \bar{y}\angle\bar{x} \vee y\angle\bar{x} \vee \bar{y}\angle x$	$y\angle x \vee \bar{y}\angle\bar{x} \vee y\angle\bar{x} \vee \bar{y}\angle x$

FUNCTIONS WITH INVOLVED CONJUNCTIONS

Venjunction and conjunction (24)

$x\angle y \vee \bar{x}\wedge\bar{y}$	$x\angle\bar{y} \vee x\wedge y$	$y\angle x \vee \bar{x}\wedge\bar{y}$	$y\angle\bar{x} \vee x\wedge y$
$x\angle y \vee x\wedge\bar{y}$	$x\angle\bar{y} \vee \bar{x}\wedge\bar{y}$	$y\angle x \vee x\wedge\bar{y}$	$y\angle\bar{x} \vee \bar{x}\wedge\bar{y}$
$x\angle y \vee \bar{x}\wedge y$	$x\angle\bar{y} \vee \bar{x}\wedge y$	$y\angle x \vee \bar{x}\wedge y$	$y\angle\bar{x} \vee x\wedge\bar{y}$
$\bar{x}\angle\bar{y} \vee x\wedge y$	$\bar{x}\angle y \vee x\wedge y$	$\bar{y}\angle\bar{x} \vee x\wedge y$	$\bar{y}\angle x \vee x\wedge y$
$\bar{x}\angle\bar{y} \vee x\wedge\bar{y}$	$\bar{x}\angle y \vee \bar{x}\wedge\bar{y}$	$\bar{y}\angle\bar{x} \vee x\wedge\bar{y}$	$\bar{y}\angle x \vee \bar{x}\wedge\bar{y}$
$\bar{x}\angle\bar{y} \vee \bar{x}\wedge y$	$\bar{x}\angle y \vee x\wedge\bar{y}$	$\bar{y}\angle\bar{x} \vee \bar{x}\wedge y$	$\bar{y}\angle x \vee \bar{x}\wedge y$

Two venjunctions and conjunction (48)

$x\angle y \vee \bar{x}\angle\bar{y} \vee x\wedge\bar{y}$	$\bar{x}\angle\bar{y} \vee x\angle\bar{y} \vee x\wedge y$	$x\angle\bar{y} \vee y\angle x \vee \bar{x}\wedge\bar{y}$
$y\angle x \vee \bar{y}\angle\bar{x} \vee x\wedge\bar{y}$	$x\angle y \vee \bar{x}\angle\bar{y} \vee \bar{x}\wedge y$	$\bar{x}\angle\bar{y} \vee x\angle\bar{y} \vee \bar{x}\wedge y$

$x \angle \bar{y} \vee y \angle x \vee \bar{x} \wedge y$	$y \angle x \vee \bar{y} \angle \bar{x} \vee \bar{x} \wedge y$	$x \angle y \vee x \angle \bar{y} \vee \bar{x} \wedge \bar{y}$
$\bar{x} \angle \bar{y} \vee \bar{x} \angle y \vee x \wedge y$	$x \angle \bar{y} \vee \bar{y} \angle \bar{x} \vee x \wedge y$	$y \angle x \vee y \angle \bar{x} \vee x \wedge \bar{y}$
$x \angle y \vee x \angle \bar{y} \vee \bar{x} \wedge y$	$\bar{x} \angle \bar{y} \vee \bar{x} \angle y \vee x \wedge \bar{y}$	$x \angle \bar{y} \vee \bar{y} \angle \bar{x} \vee \bar{x} \wedge y$
$y \angle x \vee y \angle \bar{x} \vee \bar{x} \wedge \bar{y}$	$x \angle y \vee \bar{x} \angle y \vee \bar{x} \wedge \bar{y}$	$\bar{x} \angle \bar{y} \vee y \angle x \vee x \wedge \bar{y}$
$x \angle \bar{y} \vee y \angle \bar{x} \vee x \wedge y$	$y \angle x \vee \bar{y} \angle x \vee \bar{x} \wedge \bar{y}$	$x \angle y \vee \bar{x} \angle y \vee x \wedge \bar{y}$
$\bar{x} \angle \bar{y} \vee y \angle x \vee \bar{x} \wedge y$	$x \angle \bar{y} \vee y \angle \bar{x} \vee \bar{x} \wedge \bar{y}$	$y \angle x \vee \bar{y} \angle x \vee \bar{x} \wedge y$
$x \angle y \vee \bar{y} \angle \bar{x} \vee x \wedge \bar{y}$	$\bar{x} \angle \bar{y} \vee y \angle \bar{x} \vee x \wedge y$	$\bar{x} \angle y \vee y \angle x \vee \bar{x} \wedge \bar{y}$
$\bar{y} \angle \bar{x} \vee y \angle \bar{x} \vee x \wedge y$	$x \angle y \vee \bar{y} \angle \bar{x} \vee \bar{x} \wedge y$	$\bar{x} \angle \bar{y} \vee y \angle \bar{x} \vee x \wedge \bar{y}$
$\bar{x} \angle y \vee y \angle x \vee x \wedge \bar{y}$	$\bar{y} \angle \bar{x} \vee y \angle \bar{x} \vee x \wedge \bar{y}$	$x \angle y \vee y \angle \bar{x} \vee \bar{x} \wedge \bar{y}$
$\bar{x} \angle \bar{y} \vee \bar{\bar{y}} \angle x \vee x \wedge y$	$\bar{x} \angle y \vee \bar{\bar{y}} \angle \bar{x} \vee x \wedge y$	$\bar{y} \angle \bar{x} \vee \bar{y} \angle x \vee x \wedge y$
$x \angle y \vee y \angle \bar{x} \vee x \wedge \bar{y}$	$\bar{x} \angle \bar{y} \vee \bar{y} \angle x \vee \bar{x} \wedge y$	$\bar{x} \angle y \vee \bar{y} \angle \bar{x} \vee x \wedge \bar{y}$
$\bar{y} \angle \bar{x} \vee \bar{y} \angle x \vee \bar{x} \wedge y$	$x \angle y \vee \bar{y} \angle x \vee \bar{x} \wedge \bar{y}$	$x \angle \bar{y} \vee \bar{x} \angle y \vee x \wedge y$
$\bar{x} \angle y \vee \bar{y} \angle x \vee x \wedge y$	$y \angle \bar{x} \vee \bar{y} \angle x \vee x \wedge y$	$x \angle y \vee \bar{y} \angle x \vee \bar{x} \wedge y$
$x \angle \bar{y} \vee \bar{x} \angle y \vee \bar{x} \wedge \bar{y}$	$\bar{x} \angle y \vee \bar{y} \angle x \vee \bar{x} \wedge \bar{y}$	$y \angle \bar{x} \vee \bar{y} \angle x \vee \bar{x} \wedge \bar{y}$

Three venjunctions and conjunction (32)

$x \angle y \vee \bar{x} \angle \bar{y} \vee x \angle \bar{y} \vee \bar{x} \wedge y$	$\bar{x} \angle \bar{y} \vee \bar{x} \angle y \vee \bar{y} \angle x \vee x \wedge y$
$x \angle y \vee \bar{x} \angle \bar{y} \vee \bar{x} \angle y \vee x \wedge \bar{y}$	$\bar{x} \angle \bar{y} \vee y \angle x \vee y \angle \bar{x} \vee x \wedge \bar{y}$
$x \angle y \vee x \angle \bar{y} \vee \bar{x} \angle y \vee \bar{x} \wedge \bar{y}$	$\bar{x} \angle \bar{y} \vee y \angle x \vee \bar{y} \angle x \vee \bar{x} \wedge y$
$x \angle y \vee \bar{x} \angle \bar{y} \vee y \angle \bar{x} \vee x \wedge \bar{y}$	$\bar{x} \angle \bar{y} \vee y \angle \bar{x} \vee \bar{y} \angle x \vee x \wedge y$
$x \angle y \vee \bar{x} \angle \bar{y} \vee \bar{y} \angle x \vee \bar{x} \wedge y$	$x \angle \bar{y} \vee \bar{x} \angle y \vee y \angle x \vee \bar{x} \wedge \bar{y}$
$x \angle y \vee x \angle \bar{y} \vee \bar{y} \angle \bar{x} \vee \bar{x} \wedge y$	$x \angle \bar{y} \vee \bar{x} \angle y \vee \bar{y} \angle \bar{x} \vee x \wedge y$
$x \angle y \vee x \angle \bar{y} \vee y \angle \bar{x} \vee \bar{x} \wedge \bar{y}$	$x \angle \bar{y} \vee y \angle x \vee \bar{y} \angle \bar{x} \vee \bar{x} \wedge y$
$x \angle y \vee \bar{x} \angle y \vee \bar{y} \angle \bar{x} \vee x \wedge \bar{y}$	$x \angle \bar{y} \vee y \angle x \vee y \angle \bar{x} \vee \bar{x} \wedge y$
$x \angle y \vee \bar{x} \angle y \vee \bar{y} \angle x \vee \bar{x} \wedge \bar{y}$	$x \angle \bar{y} \vee \bar{y} \angle \bar{x} \vee y \angle \bar{x} \vee x \wedge y$

$x \angle y \vee \overline{y} \angle \overline{x} \vee y \angle \overline{x} \vee x \wedge \overline{y}$	$\overline{x} \angle y \vee y \angle x \vee \overline{y} \angle \overline{x} \vee x \wedge \overline{y}$
$x \angle y \vee \overline{y} \angle \overline{x} \vee \overline{y} \angle x \vee \overline{x} \wedge y$	$\overline{x} \angle y \vee y \angle x \vee \overline{y} \angle x \vee \overline{x} \wedge \overline{y}$
$x \angle y \vee y \angle \overline{x} \vee \overline{y} \angle x \vee \overline{x} \wedge \overline{y}$	$\overline{x} \angle y \vee \overline{y} \angle \overline{x} \vee \overline{y} \angle x \vee x \wedge y$
$\overline{x} \angle \overline{y} \vee x \angle \overline{y} \vee \overline{x} \angle y \vee x \wedge y$	$y \angle x \vee \overline{y} \angle \overline{x} \vee y \angle \overline{x} \vee x \wedge \overline{y}$
$\overline{x} \angle \overline{y} \vee x \angle \overline{y} \vee y \angle x \vee \overline{x} \wedge y$	$y \angle x \vee \overline{y} \angle \overline{x} \vee \overline{y} \angle x \vee \overline{x} \wedge y$
$\overline{x} \angle \overline{y} \vee x \angle \overline{y} \vee y \angle \overline{x} \vee x \wedge y$	$y \angle x \vee y \angle \overline{x} \vee \overline{y} \angle x \vee \overline{x} \wedge \overline{y}$
$\overline{x} \angle \overline{y} \vee \overline{x} \angle y \vee y \angle x \vee x \wedge \overline{y}$	$\overline{y} \angle \overline{x} \vee y \angle \overline{x} \vee \overline{y} \angle x \vee x \wedge y$

Venjunction and two conjunctions (8)

$x \angle y \vee x \wedge \overline{y} \vee \overline{x} \wedge y$	$x \angle \overline{y} \vee x \wedge y \vee \overline{x} \wedge \overline{y}$	$y \angle x \vee x \wedge \overline{y} \vee \overline{x} \wedge y$
$y \angle \overline{x} \vee x \wedge y \vee \overline{x} \wedge \overline{y}$	$\overline{x} \angle \overline{y} \vee x \wedge \overline{y} \vee \overline{x} \wedge y$	$\overline{x} \angle y \vee x \wedge y \vee \overline{x} \wedge \overline{y}$
$\overline{y} \angle \overline{x} \vee x \wedge \overline{y} \vee \overline{x} \wedge y$	$\overline{y} \angle x \vee x \wedge y \vee \overline{x} \wedge \overline{y}$	-

Two venjunctions and two conjunctions (8)

$x \angle y \vee \overline{x} \angle \overline{y} \vee x \wedge \overline{y} \vee \overline{x} \wedge y$	$x \angle \overline{y} \vee y \angle \overline{x} \vee x \wedge y \vee \overline{x} \wedge \overline{y}$
$x \angle y \vee \overline{y} \angle \overline{x} \vee x \wedge \overline{y} \vee \overline{x} \wedge y$	$\overline{x} \angle y \vee \overline{y} \angle x \vee x \wedge y \vee \overline{x} \wedge \overline{y}$
$\overline{x} \angle \overline{y} \vee y \angle x \vee x \wedge \overline{y} \vee \overline{x} \wedge y$	$y \angle x \vee \overline{y} \angle \overline{x} \vee x \wedge \overline{y} \vee \overline{x} \wedge y$
$x \angle \overline{y} \vee \overline{x} \angle y \vee x \wedge y \vee \overline{x} \wedge \overline{y}$	$y \angle \overline{x} \vee \overline{y} \angle x \vee x \wedge y \vee \overline{x} \wedge \overline{y}$

FUNCTIONS WITH INVOLVING VARIABLES

Venjunction and variable (16)

$x \angle y \vee \overline{x}$	$x \angle \overline{y} \vee \overline{x}$	$y \angle x \vee \overline{x}$	$y \angle \overline{x} \vee x$
$x \angle y \vee \overline{y}$	$x \angle \overline{y} \vee y$	$y \angle x \vee \overline{y}$	$y \angle \overline{x} \vee \overline{y}$
$\overline{x} \angle \overline{y} \vee x$	$\overline{x} \angle y \vee x$	$\overline{y} \angle \overline{x} \vee x$	$\overline{y} \angle x \vee \overline{x}$
$\overline{x} \angle \overline{y} \vee y$	$\overline{x} \angle y \vee \overline{y}$	$\overline{y} \angle \overline{x} \vee y$	$\overline{y} \angle x \vee y$

Two venjunctions and variable (16)		
$x \angle y \vee x \angle \overline{y} \vee \overline{x}$	$\overline{x} \angle \overline{y} \vee x \angle \overline{y} \vee y$	$x \angle \overline{y} \vee y \angle x \vee \overline{x}$
$y \angle x \vee y \angle \overline{x} \vee \overline{y}$	$x \angle y \vee \overline{x} \angle y \vee \overline{y}$	$\overline{x} \angle \overline{y} \vee \overline{x} \angle y \vee x$
$x \angle \overline{y} \vee \overline{y} \angle \overline{x} \vee y$	$y \angle x \vee \overline{y} \angle x \vee \overline{x}$	$x \angle y \vee y \angle \overline{x} \vee \overline{y}$
$\overline{x} \angle \overline{y} \vee y \angle \overline{x} \vee x$	$\overline{x} \angle y \vee y \angle x \vee \overline{y}$	$\overline{y} \angle \overline{x} \vee y \angle \overline{x} \vee x$
$x \angle y \vee \overline{y} \angle x \vee \overline{x}$	$\overline{x} \angle \overline{y} \vee \overline{y} \angle x \vee y$	$\overline{x} \angle y \vee \overline{y} \angle \overline{x} \vee x$
$\overline{y} \angle \overline{x} \vee \overline{y} \angle x \vee y$	-	-
Venjunction and two variables (8)		
$x \angle y \vee \overline{x} \vee \overline{y}$	$x \angle \overline{y} \vee \overline{x} \vee y$	$y \angle x \vee \overline{x} \vee \overline{y}$
$y \angle \overline{x} \vee x \vee \overline{y}$	$\overline{x} \angle \overline{y} \vee x \vee y$	$\overline{x} \angle y \vee x \vee \overline{y}$
$\overline{y} \angle \overline{x} \vee x \vee y$	$\overline{y} \angle x \vee \overline{x} \vee y$	-

Appendix B: The List of Typical Trigger Functions

X	Y	Φ
1 venjunction	1 venjunction	
$x \angle y$	$y \angle x$	$\overline{x} \vee \overline{y}$
$x \angle y$	$x \angle \overline{y}$	$y \angle x \vee \overline{y} \angle x \vee \overline{x}$
$x \angle y$	$\overline{x} \angle \overline{y}$	$y \angle x \vee \overline{y} \angle \overline{x} \vee x \wedge \overline{y} \vee \overline{x} \wedge y$
1 venjunction	2 venjunctions	
$x \angle y$	$y \angle x \vee \overline{y} \angle x$	$x \angle \overline{y} \vee \overline{x}$
$x \angle y$	$y \angle x \vee \overline{y} \angle \overline{x}$	$\overline{x} \angle \overline{y} \vee x \wedge \overline{y} \vee \overline{x} \wedge y$
$x \angle y$	$x \angle \overline{y} \vee \overline{x} \angle \overline{y}$	$y \angle x \vee \overline{y} \angle x \vee \overline{y} \angle \overline{x} \vee \overline{x} \wedge y$
1 venjunction	3 venjunctions	
$x \angle y$	$y \angle x \vee \overline{y} \angle x \vee \overline{y} \angle \overline{x}$	$x \angle \overline{y} \vee x \angle \overline{y} \vee \overline{x} \wedge y$
$x \angle y$	$x \angle \overline{y} \vee \overline{x} \angle \overline{y} \vee \overline{x} \angle y$	$y \angle x \vee \overline{y} \angle x \vee \overline{y} \angle \overline{x} \vee y \angle \overline{x}$
1 venjunction	4 venjunctions	
$x \angle y$	$y \angle x \vee \overline{y} \angle x \vee \overline{y} \angle \overline{x} \vee y \angle \overline{x}$	$x \angle \overline{y} \vee \overline{x} \angle \overline{y} \vee \overline{x} \angle y$
2 venjunctions	2 venjunctions	
$x \angle y \vee \overline{x} \angle \overline{y}$	$y \angle x \vee \overline{y} \angle \overline{x}$	$x \wedge \overline{y} \vee \overline{x} \wedge y$
$x \angle y \vee x \angle \overline{y}$	$y \angle x \vee \overline{y} \angle x$	$\overline{x}$
$x \angle y \vee x \angle \overline{y}$	$\overline{y} \angle x \vee \overline{y} \angle \overline{x}$	$y \angle x \vee \overline{x} \angle \overline{y} \vee \overline{x} \wedge y$
$x \angle y \vee x \angle \overline{y}$	$\overline{x} \angle \overline{y} \vee \overline{x} \angle y$	$y \angle x \vee \overline{y} \angle x \vee \overline{y} \angle \overline{x} \vee y \angle \overline{x}$

2 venjunctions	3 venjunctions	
$x\angle y \vee x\angle \overline{y}$	$y\angle x \vee \overline{y}\angle x \vee \overline{y}\angle \overline{x}$	$\overline{x}\angle \overline{y} \vee \overline{x}\wedge y$
$x\angle y \vee x\angle \overline{y}$	$\overline{y}\angle x \vee \overline{y}\angle \overline{x} \vee y\angle \overline{x}$	$y\angle x \vee \overline{x}\angle \overline{y} \vee \overline{x}\angle y$
2 venjunctions	**4 venjunctions**	
$x\angle y \vee x\angle \overline{y}$	$y\angle x \vee \overline{y}\angle x \vee \overline{y}\angle \overline{x} \vee y\angle \overline{x}$	$\overline{x}\angle \overline{y} \vee \overline{x}\angle y$
3 venjunctions	**3 venjunctions**	
$x\angle y \vee x\angle \overline{y} \vee \overline{x}\angle \overline{y}$	$y\angle x \vee \overline{y}\angle x \vee \overline{y}\angle \overline{x}$	$\overline{x}\wedge y$
$x\angle y \vee x\angle \overline{y} \vee \overline{x}\angle \overline{y}$	$\overline{y}\angle x \vee \overline{y}\angle \overline{x} \vee y\angle \overline{x}$	$y\angle x \vee \overline{x}\angle y$
3 venjunctions	**4 venjunctions**	
$x\angle y \vee x\angle \overline{y} \vee \overline{x}\angle \overline{y}$	$y\angle x \vee \overline{y}\angle x \vee \overline{y}\angle \overline{x} \vee y\angle \overline{x}$	$\overline{x}\wedge y$
1 venjunction	**1 conjunction**	
$x\angle y$	$x\wedge \overline{y}$	$y\angle x \vee \overline{x}$
$x\angle y$	$\overline{x}\wedge \overline{y}$	$y\angle x \vee x\wedge \overline{y} \vee \overline{x}\wedge y$
2 venjunctions	**1 conjunction**	
$x\angle y \vee x\angle \overline{y}$	$\overline{x}\wedge \overline{y}$	$y\angle x \vee \overline{y}\angle x \vee \overline{x}\wedge y$
3 venjunctions	**1 conjunction**	
$x\angle y \vee x\angle \overline{y} \vee \overline{x}\angle \overline{y}$	$\overline{x}\wedge y$	$y\angle x \vee \overline{y}\angle x \vee \overline{y}\angle \overline{x}$
1 venjunction	**1 variable**	
$x\angle y$	$\overline{x}$	$y\angle x \vee x\wedge \overline{y}$
2 venjunctions	**1 variable**	
$x\angle y \vee x\angle \overline{y}$	$\overline{x}$	$y\angle x \vee \overline{y}\angle x$

1 conjunction	1 conjunction	
$x \wedge y$	$x \wedge \overline{y}$	$\overline{x}$
$x \wedge y$	$\overline{x} \wedge \overline{y}$	$x \wedge \overline{y} \vee \overline{x} \wedge y$
1 conjunction	1 variable	
$x \wedge y$	$\overline{x}$	$x \wedge \overline{y}$

Appendix C: Examples of Trigger-Type Devices

Example 1

Trigger, output settings of which are caused by signal switchings formed on the basis of venjunctions of initial variables.

- Unity setting: $X = (x \angle y) \vee (\overline{y} \angle x)$.
- Zero setting: $Y = (\overline{y} \angle \overline{x})$.
- Function: $Z = x \angle y \vee \overline{y} \angle x \vee \overline{(x \angle y \vee \overline{y} \angle x)} \angle \overline{(\overline{y} \angle \overline{x})}$.
- Memory formula: $\Phi = \overline{x} \wedge y \vee x \angle \overline{y} \vee \overline{x} \angle \overline{y} \vee y \angle x$.

Cycles of switchings:

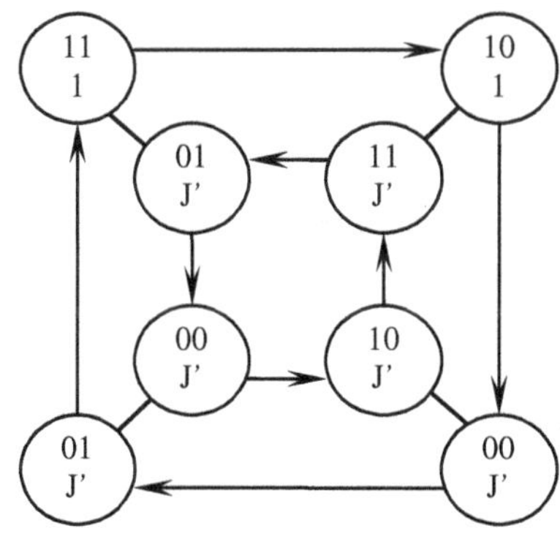

Example 2

Trigger with a pair of mirror venjunctions intended for setting actions.

- Unity setting: $X = (x \angle y)$.
- Zero setting: $Y = (y \angle x)$.
- Function: $Z = x \angle y \vee \overline{(x \angle y)} \angle \overline{(y \angle x)}$.
- Memory formula: $\Phi = \overline{x} \vee \overline{y}$.

Cycles of switchings:

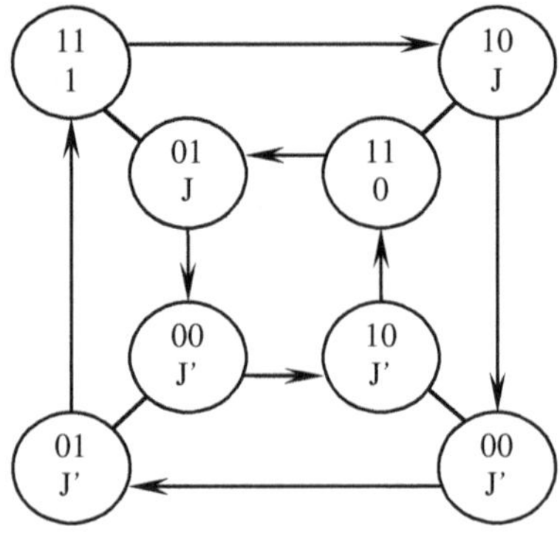

Example 3

Trigger with two pairs of mirror venjunctions intended for setting actions.

- Unity setting: $X = (x \angle y) \vee (\overline{x} \angle \overline{y})$.
- Zero setting: $Y = (y \angle x) \vee (\overline{y} \angle \overline{x})$.
- Function: $Z = x \angle y \vee \overline{x} \angle \overline{y} \vee \overline{(x \angle y \vee \overline{x} \angle \overline{y})} \angle \overline{(y \angle x \vee \overline{y} \angle \overline{x})}$.
- Memory formula: $\Phi = x \wedge \overline{y} \vee \overline{x} \wedge y = x \oplus y$.

Cycles of switchings:

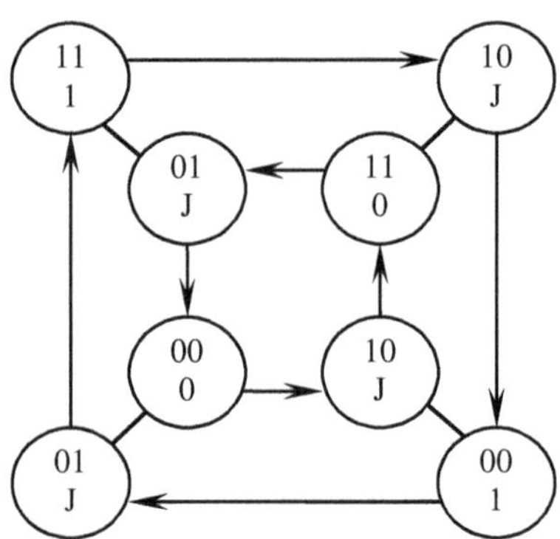

Example 4

Trigger with settings caused by three pairs of mirror venjunctions.

- Unity setting: $X = (x \angle y) \vee (\overline{x} \angle \overline{y}) \vee (x \angle \overline{y})$.
- Zero setting: $Y = (y \angle x) \vee (\overline{y} \angle \overline{x}) \vee (\overline{y} \angle x)$.

- Function:

$$Z = x \angle y \vee \overline{x} \angle \overline{y} \vee x \angle \overline{y} \vee \overline{(x \angle y \vee \overline{x} \angle \overline{y} \vee x \angle \overline{y})} \angle \overline{(y \angle x \vee \overline{y} \angle \overline{x} \vee \overline{y} \angle x)}.$$

- Memory formula: $\Phi = \overline{x} \wedge y$.

Cycles of switchings:

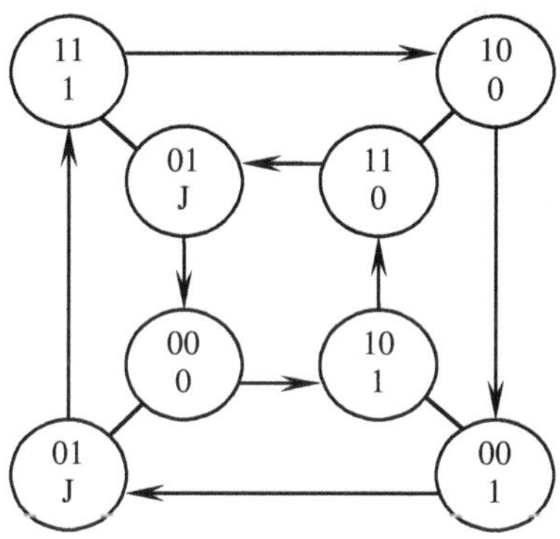

Example 5

Trigger with a venjunction intended for memory state.

- Unity setting: $X = y \vee \overline{x} \angle \overline{y}$.
- Zero setting: $Y = \overline{y} \angle x \vee \overline{y} \angle \overline{x}$.
- Function: $Z = y \vee \overline{x} \angle \overline{y} \vee \overline{(y \vee \overline{x} \angle \overline{y})} \angle \overline{(\overline{y} \angle x \vee \overline{y} \angle \overline{x})}$.
- Memory formula: $\Phi = (x \angle \overline{y})$.

Cycles of switchings:

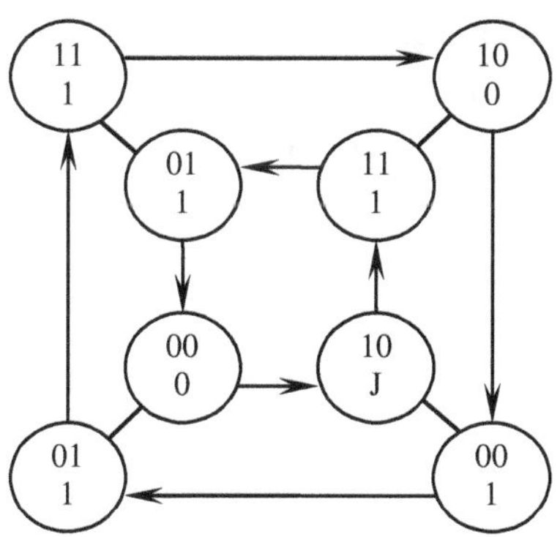

Example 6

Trigger with two venjunctions intended for memory state.

- Unity setting: $X = x \angle y \vee x \angle \overline{y} \vee y \angle \overline{x} \vee \overline{y} \angle \overline{x}$.
- Zero setting: $Y = y \angle x \vee \overline{y} \angle x$.
- Function:

$$Z = x \angle y \vee x \angle \overline{y} \vee y \angle \overline{x} \vee \overline{y} \angle \overline{x} \vee \overline{(x \angle y \vee x \angle \overline{y} \vee y \angle \overline{x} \vee \overline{y} \angle \overline{x})} \angle \overline{(y \angle x \vee \overline{y} \angle x)} .$$

- Memory formula: $\Phi = (\overline{x} \angle y) \vee (\overline{x} \angle \overline{y})$.

Cycles of switchings:

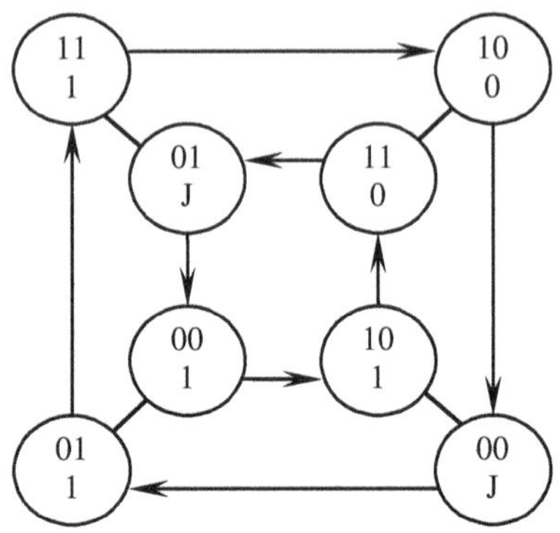

Example 7

Trigger with three venjunctions intended for memory.

- Unity setting: $X = x \angle y \vee \overline{x} \angle \overline{y}$.
- Zero setting: $Y = \overline{y} \angle x \vee \overline{x} \angle y \vee y \angle x$
- Function: $Z = x \angle y \vee \overline{x} \angle \overline{y} \vee \overline{(x \angle y \vee \overline{x} \angle \overline{y})} \angle \overline{(\overline{y} \angle x \vee \overline{x} \angle y \vee y \angle x)}$.
- Memory formula: $\Phi = (x \angle \overline{y}) \vee (y \angle \overline{x}) \vee (\overline{y} \angle \overline{x})$.

Cycles of switchings:

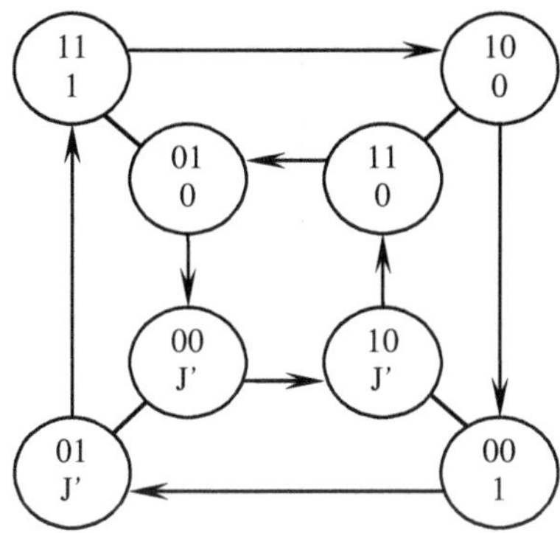

Example 8

Trigger with four venjunctions intended for memory.

- Unity setting: $X = x\angle y \vee \overline{x} \angle \overline{y}$.
- Zero setting: $Y = y\angle \overline{x} \vee \overline{y} \angle x$.
- Function: $Z = x\angle y \vee \overline{x} \angle \overline{y} \vee \overline{(x\angle y \vee \overline{x} \angle \overline{y})} \angle \overline{(y\angle \overline{x} \vee \overline{y} \angle x)}$.
- Memory formula: $\Phi = (x\angle \overline{y}) \vee (\overline{x} \angle y) \vee (y\angle x) \vee (\overline{y} \angle \overline{x})$.

Cycles of switchings:

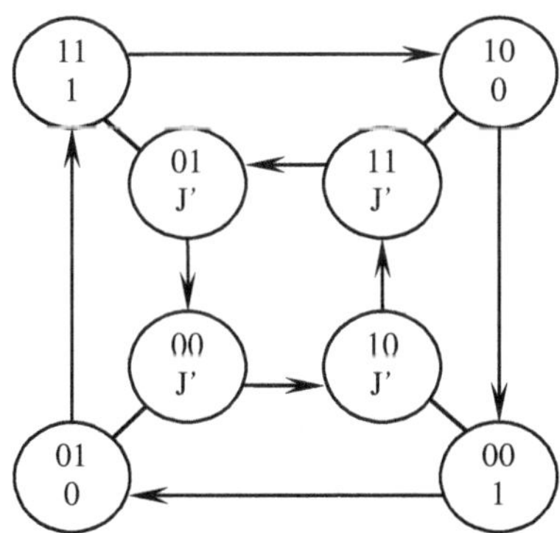

Appendix D: Asynchronous Sequential Circuits

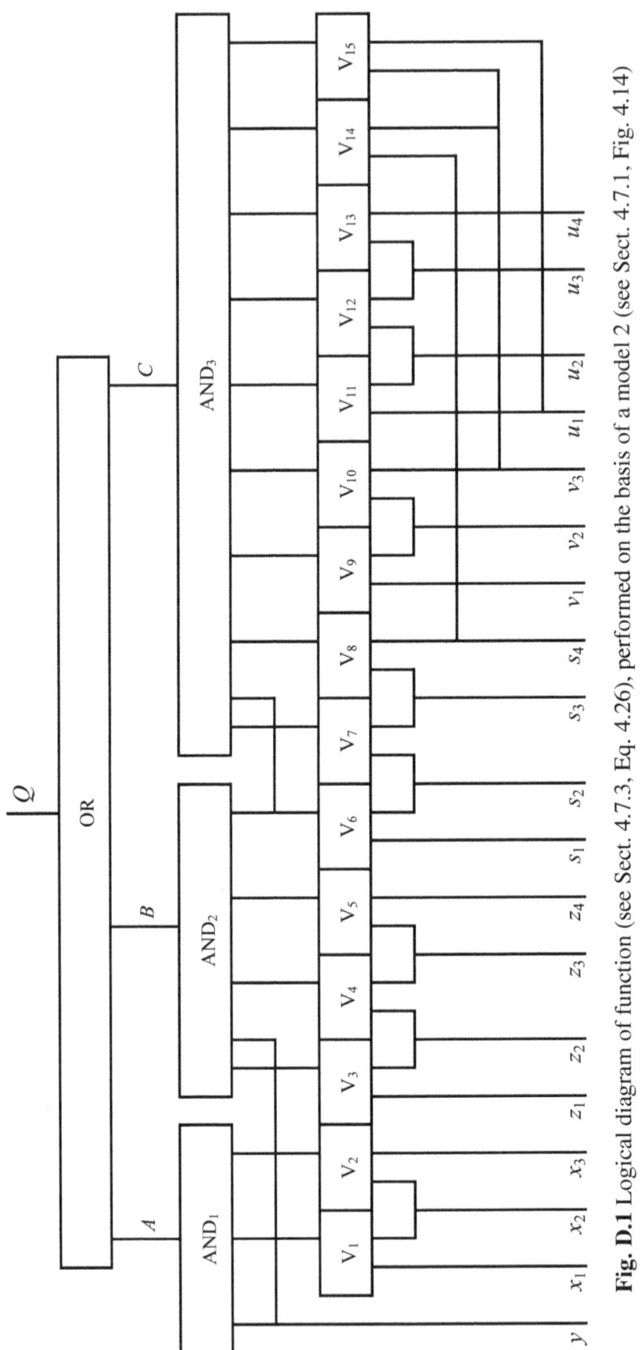

Fig. D.1 Logical diagram of function (see Sect. 4.7.3, Eq. 4.26), performed on the basis of a model 2 (see Sect. 4.7.1, Fig. 4.14)

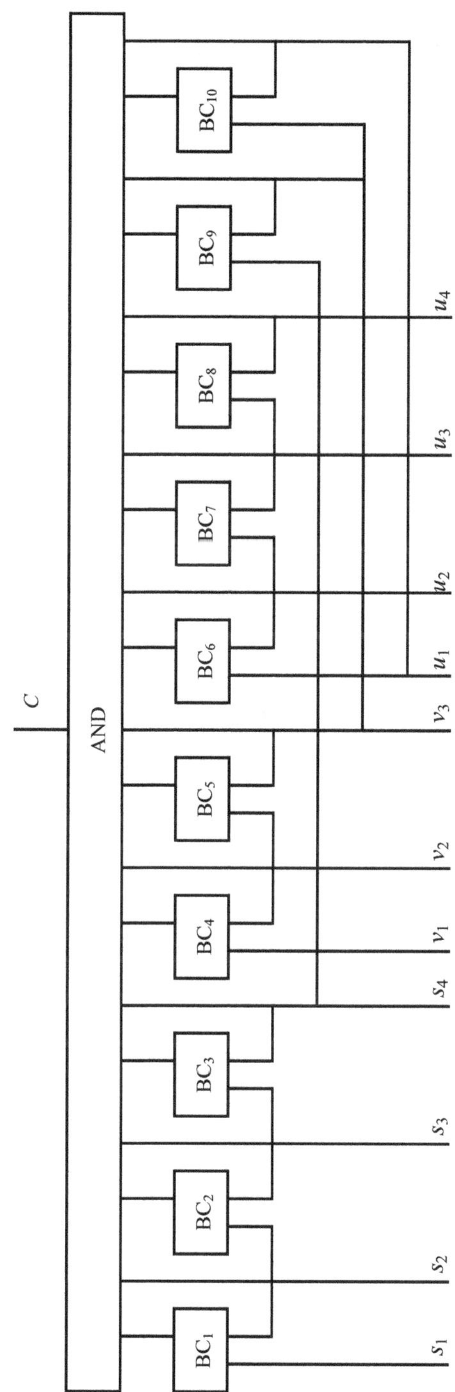

Fig. D.2 Logical diagram of subfunction (see Sect. 4.7.3, Eq. 4.28), performed on the basis of a model 1 (see Sect. 4.7.1, Fig. 4.14)